Dimensions Math®
Workbook KA

Authors and Reviewers

Pearly Yuen

Tricia Salerno

Jenny Kempe

Allison Coates

Singapore Math Inc.

Published by Singapore Math Inc.

19535 SW 129th Avenue
Tualatin, OR 97062
www.singaporemath.com

Dimensions Math® Workbook Kindergarten A
ISBN 978-1-947226-16-6

First published 2018
Reprinted 2018, 2019 (twice)

Printed in China

Acknowledgments

Editing by the Singapore Math Inc. team.
Design and illustration by Cameron Wray with Carli Fronius.

Contents

Blank

Chapter 1 Match, Sort, and Classify

Exercise 1

Circle the animals facing left.

Using this page: Have students identify the animals that are facing the left side of the page and circle.
Concept: Identifying left.

Color the things on the left.

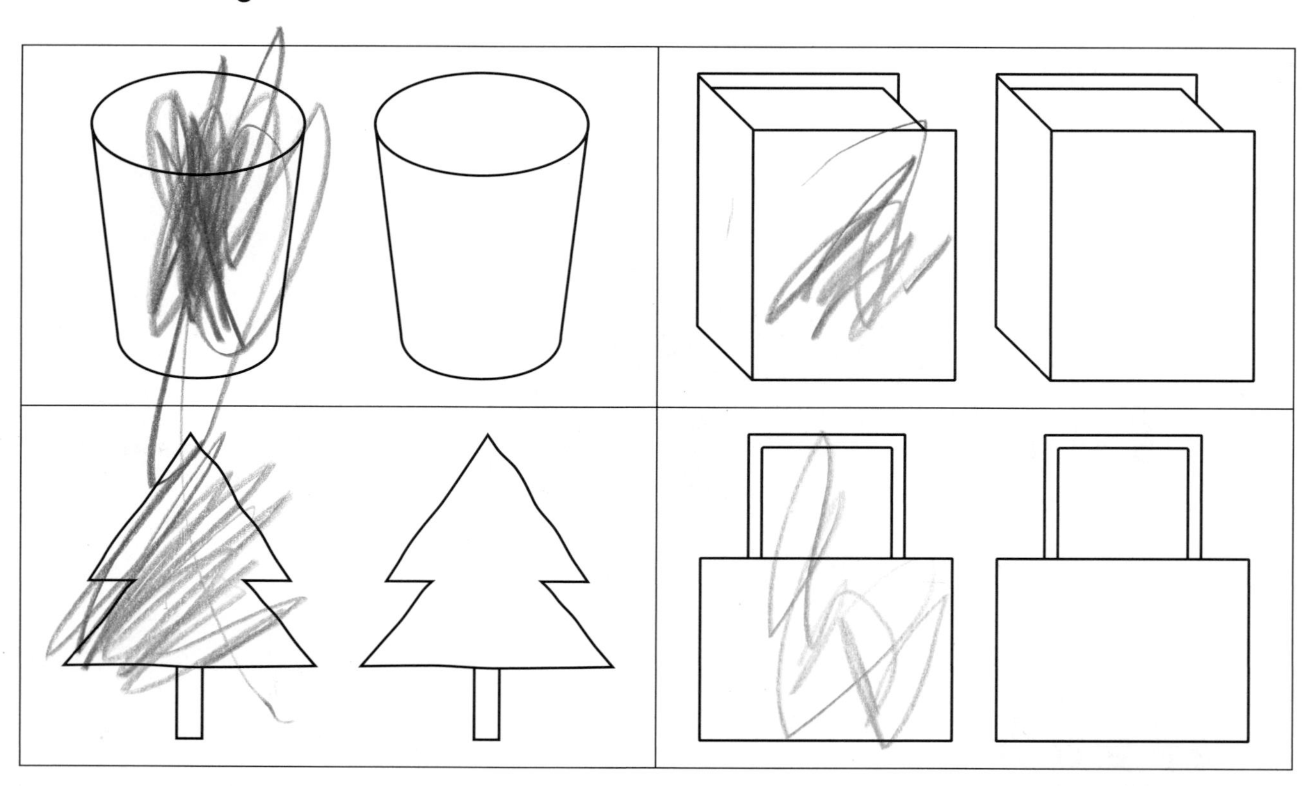

Color the things on the right.

Using this page: Have students color the objects as specified.
Concept: Distinguishing left and right.

Exercise 2

Circle the 2 penguins that are the same.

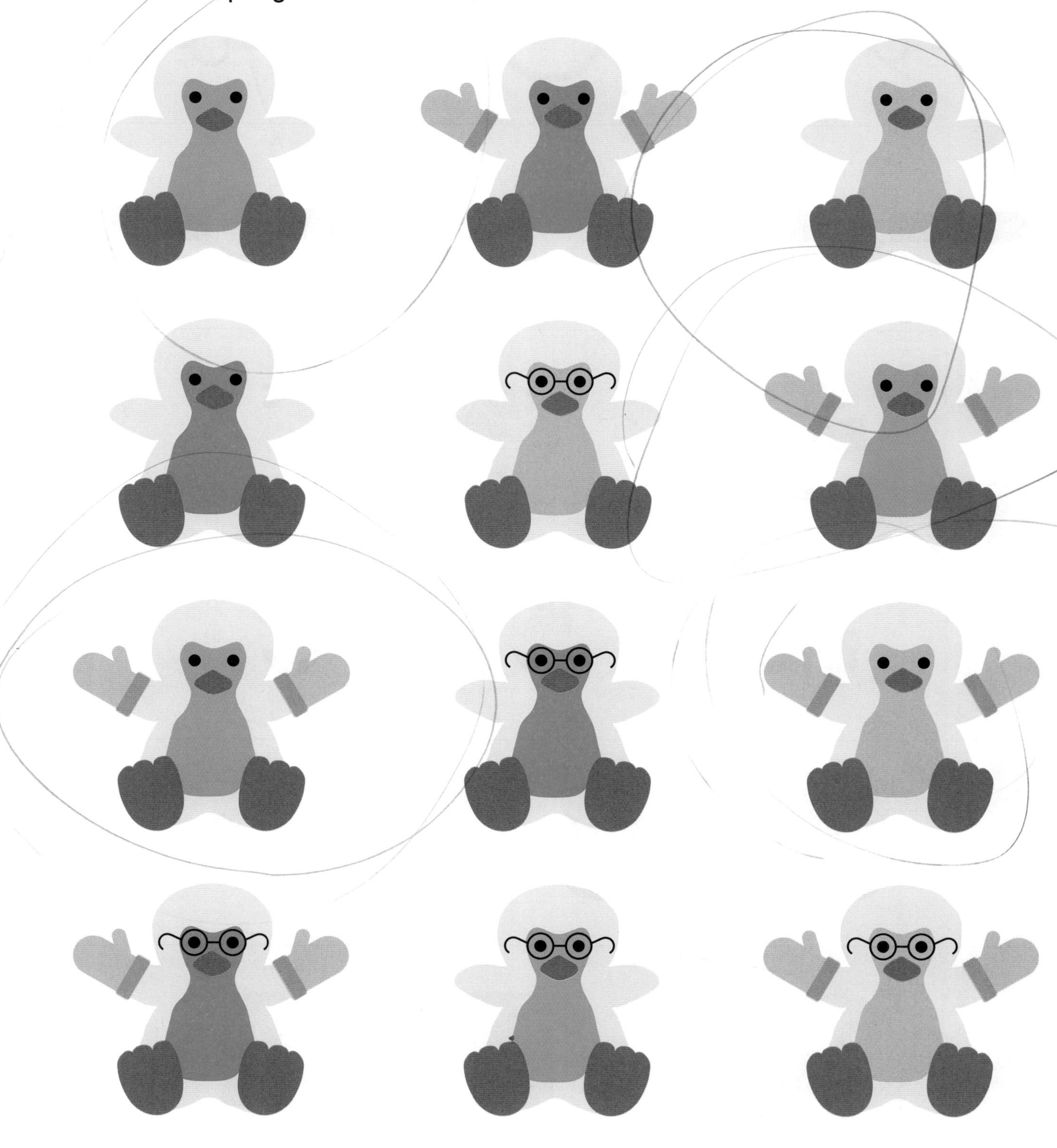

Using this page: Have students look for the two penguins that are exactly the same and circle them.
Concept: Identifying objects that are the same.

Circle the things that are similar.

Using this page: Have students circle the objects that are similar.
Concept: Identifying objects that are similar.

Exercise 3

Cross out the thing that is different.

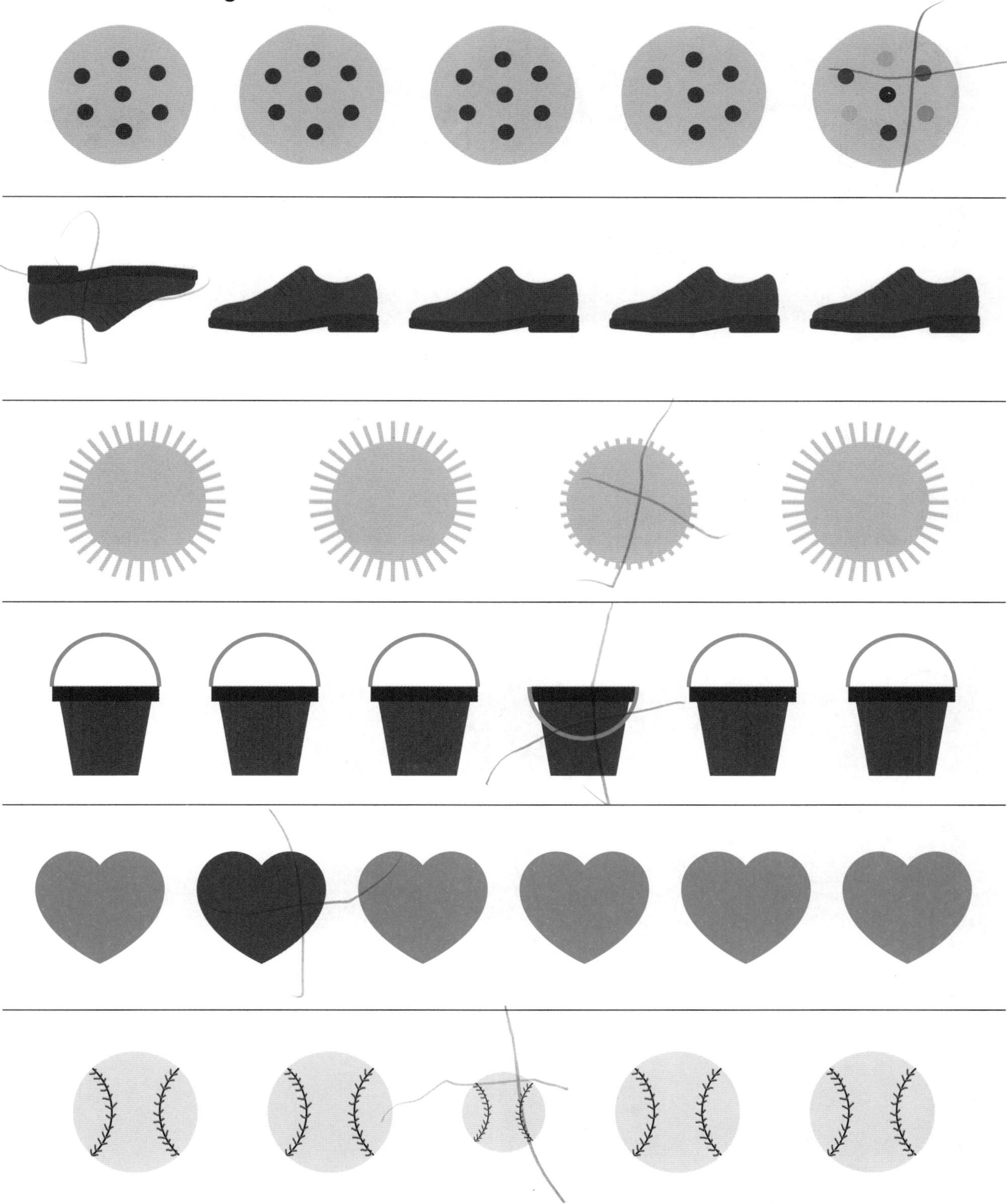

Using this page: Have students cross out the object that is different in each row.
Concept: Identifying the object that is different.

Circle the 3 things that are different in the bottom picture.

Using this page: Have students look for the three differences in the bottom picture and circle.
Concept: Identifying what is different.

Exercise 4

Circle the things that are rough.

Using this page: Have students identify objects that are rough, and circle.
Concept: Identifying rough objects.

Circle the things that are smooth.

Using this page: Have students identify the objects that are smooth, and circle.
Concept: Identifying smooth objects.

Exercise 5

Match.

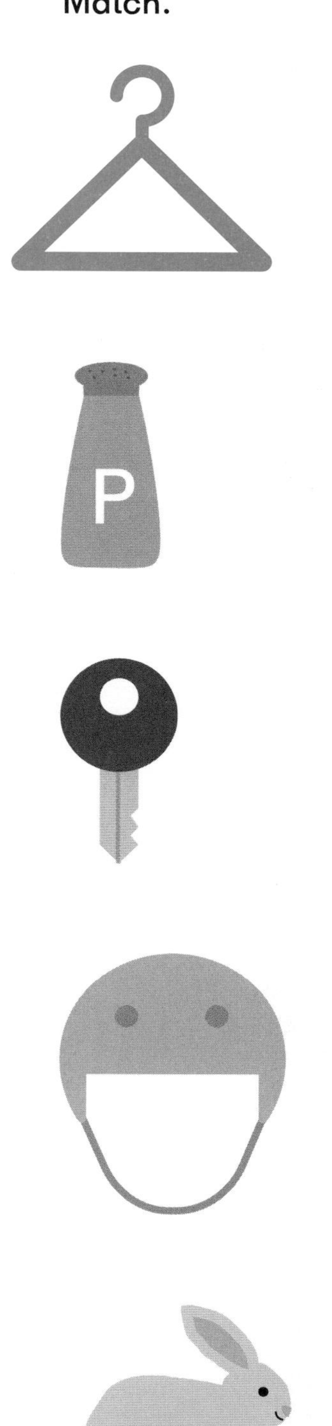

Using this page: Have students match the things that go together.
Concept: Identifying objects that go together.

Circle the things that go together.

Using this page: Have students identify the objects that go together and circle them.
Concept: Matching objects that go together.

Exercise 6

Sort the buttons.

Before using this page: Pre-cut the buttons on page 145.
Using this page: Have students sort the buttons into two or more groups of their choice (for example, sorting by color, shape, or number of button holes), then paste them in the boxes.
Concept: Sorting by a selected criterion.

Sort the flowers.

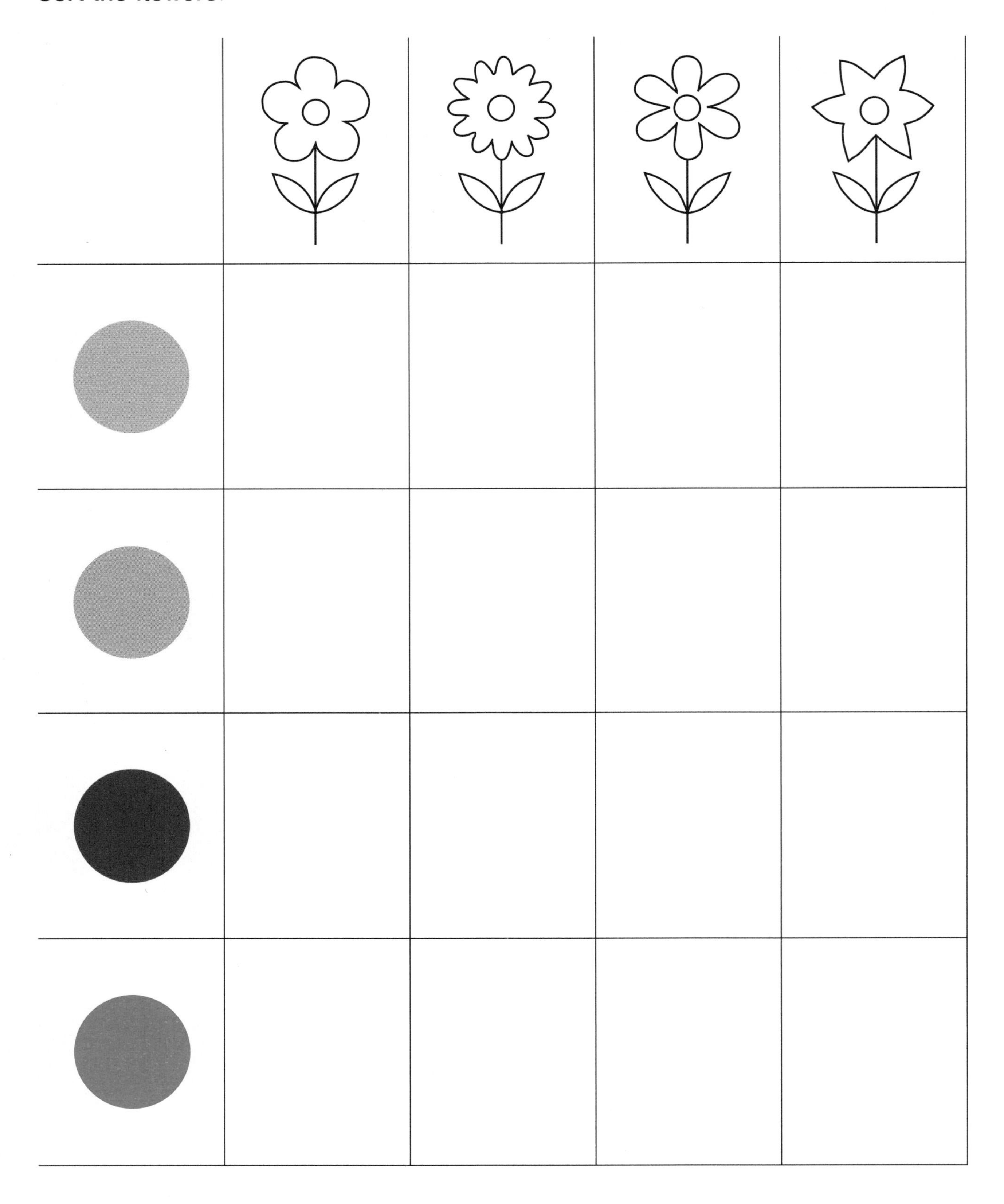

Before using this page: Pre-cut the flowers on page 145.
Using this page: Have students sort the flowers by their types, then match them to the correct color row and paste.
Concept: Sorting by a given criterion.

Exercise 7

Color to match the socks.

Using this page: Have students look for two socks that are the same and color them the same color.

Cross out the thing that does not go with the others.

Using this page: Have students determine how to classify the objects and cross out the one that does not belong.

Chapter 2 Numbers to 5

Exercise 1

Match.

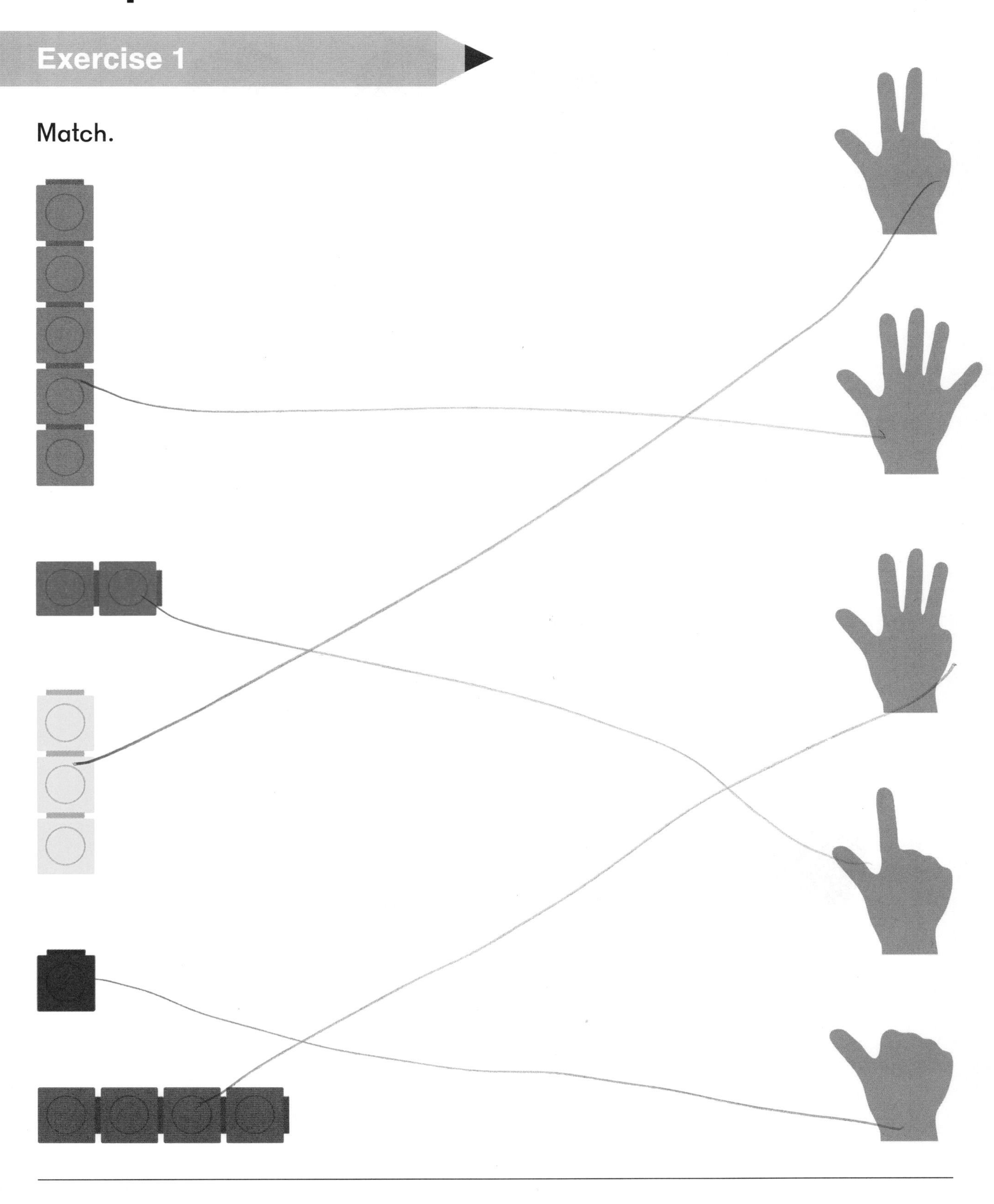

Using this page: Have students count the linking cubes and match them with the same number of fingers shown.
Concept: One-to-one correspondence and cardinality.

Circle the groups of 4.

Using this page: Have students count the objects and circle the groups of four.
Concept: One-to-one correspondence and cardinality.

Exercise 2

Match.

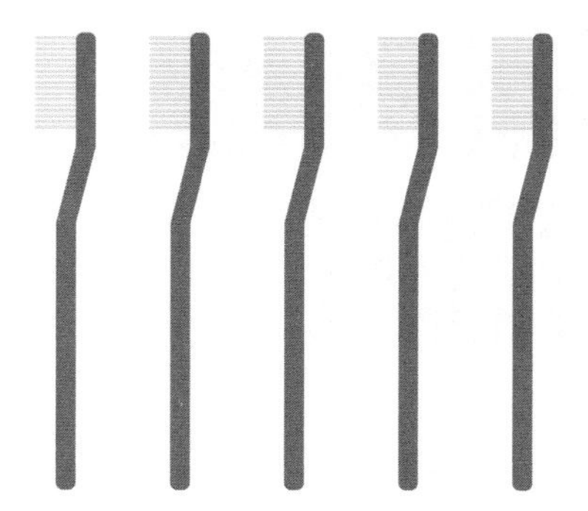

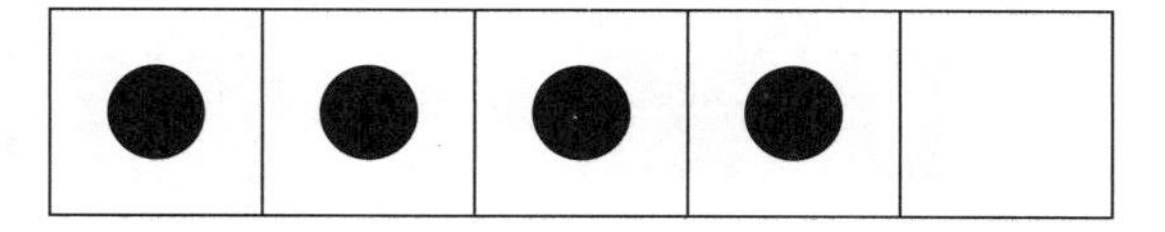

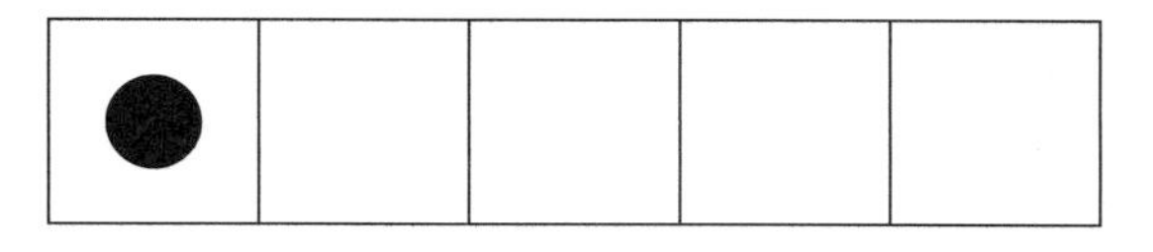

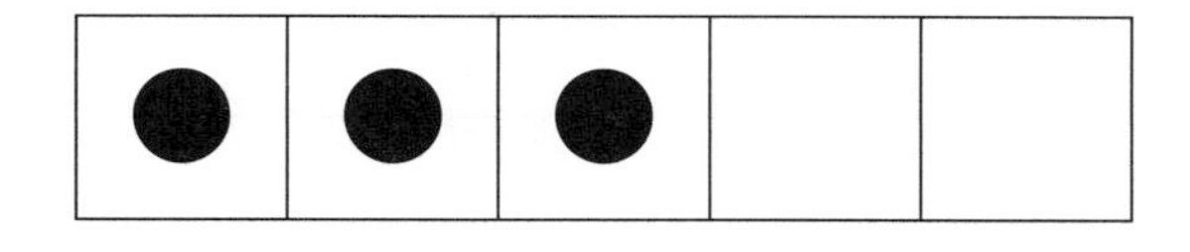

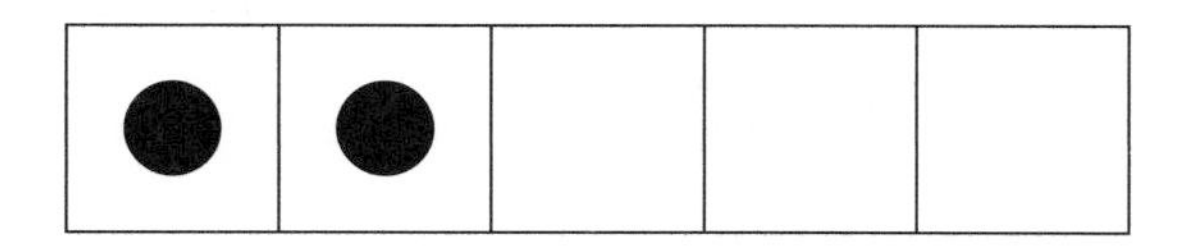

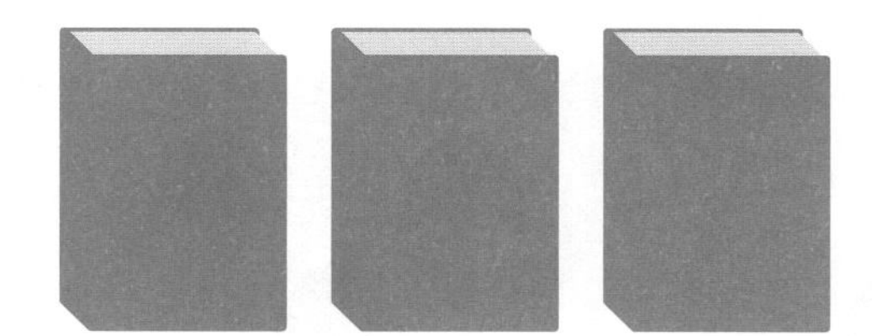

Using this page: Have students count and cross out the objects to help them keep track if needed, then match with the five-frame card.
Concept: One-to-one correspondence and cardinality.

Circle the group that has a different number of things.

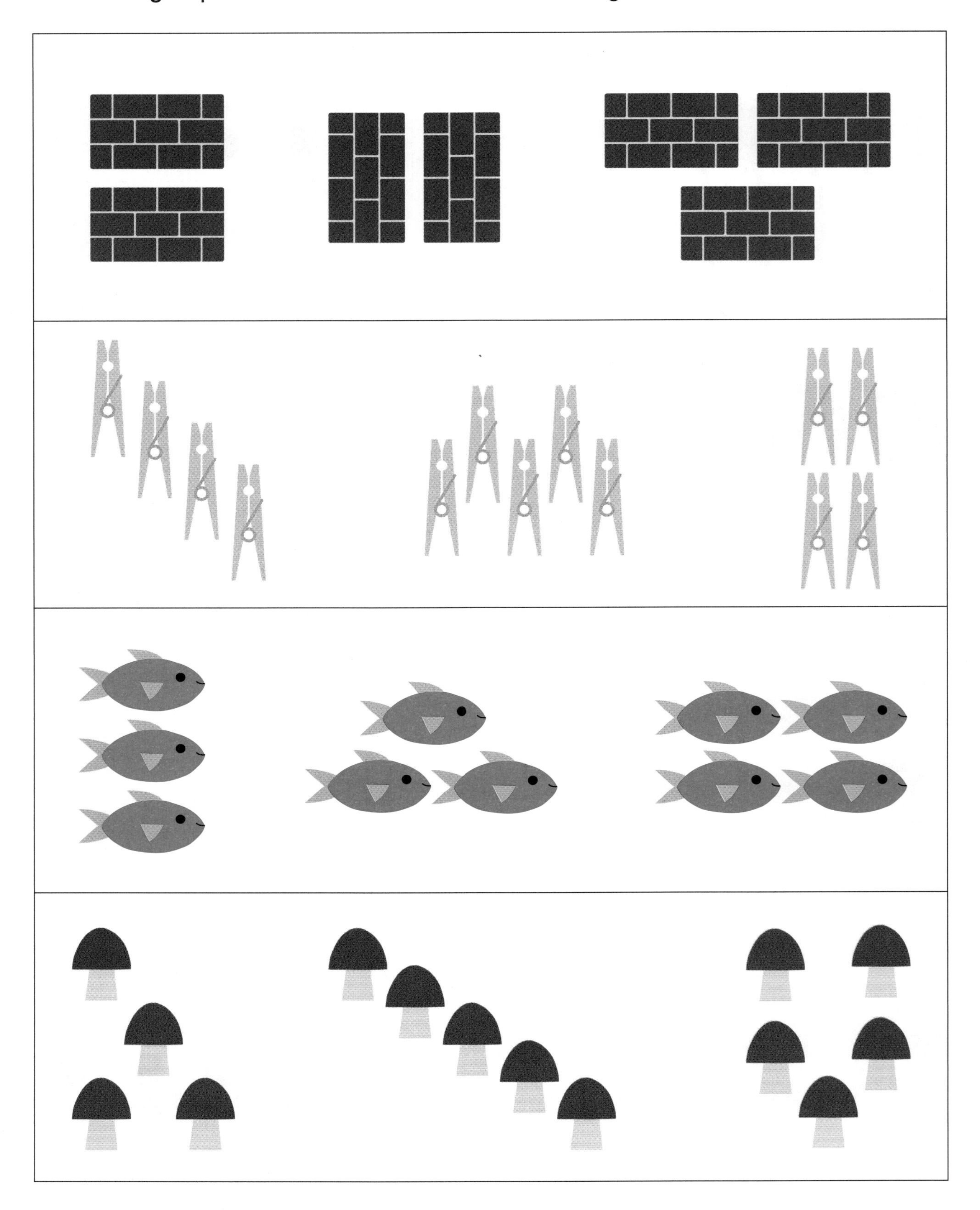

Using this page: Have students count and circle the group of objects that has a different number in each row.
Concept: One-to-one correspondence and cardinality.

Exercise 3

Match.

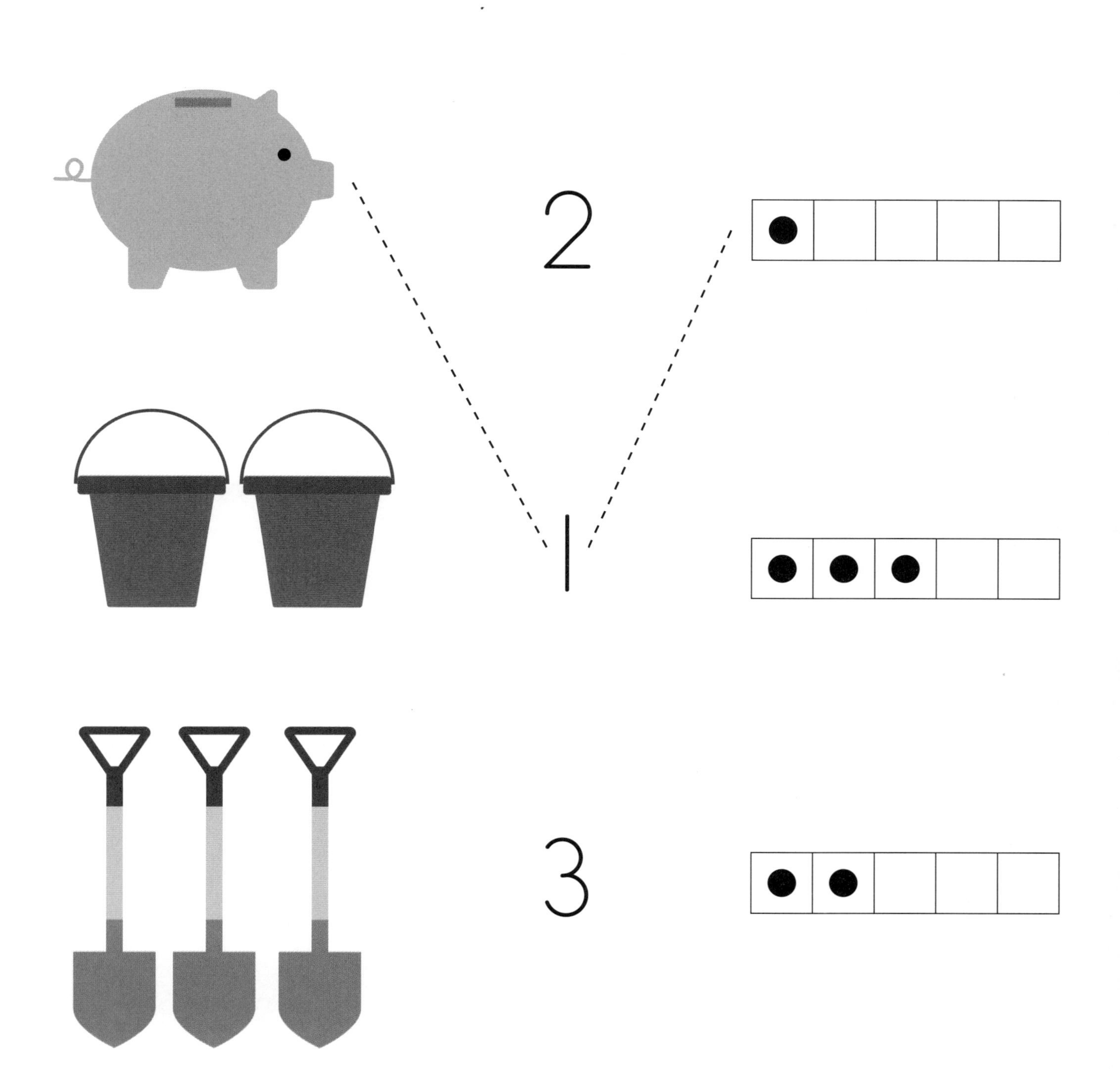

Using this page: Have students match each numeral to the correct set of objects and five-frame.
Concept: Numeral recognition 1 to 3.

Follow the directions and draw.

Draw 2 balls.

Draw 3 stars.

Draw 1 thing that you like.

Using this page: Direct students to follow the directions and draw the specified number of objects.
Concept: Numeral recognition 1 to 3 and cardinality.

Exercise 4

Find and color.

- 4 🐸
- 5 🐤

Using this page: Have students search for the number of animals as specified in the legend and color them.
Concept: Numeral recognition 4 and 5.

Color according to the Color Key.

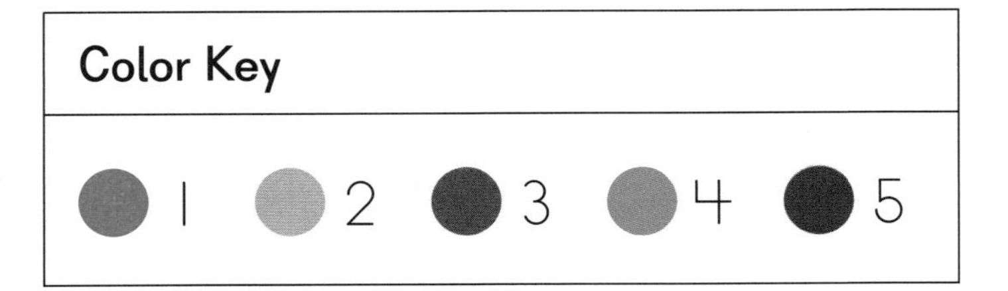

Using this page: Have students refer to the color key and color the picture.
Concept: Numeral recognition 1 to 5.

Exercise 5

Circle the cubes that match the number.

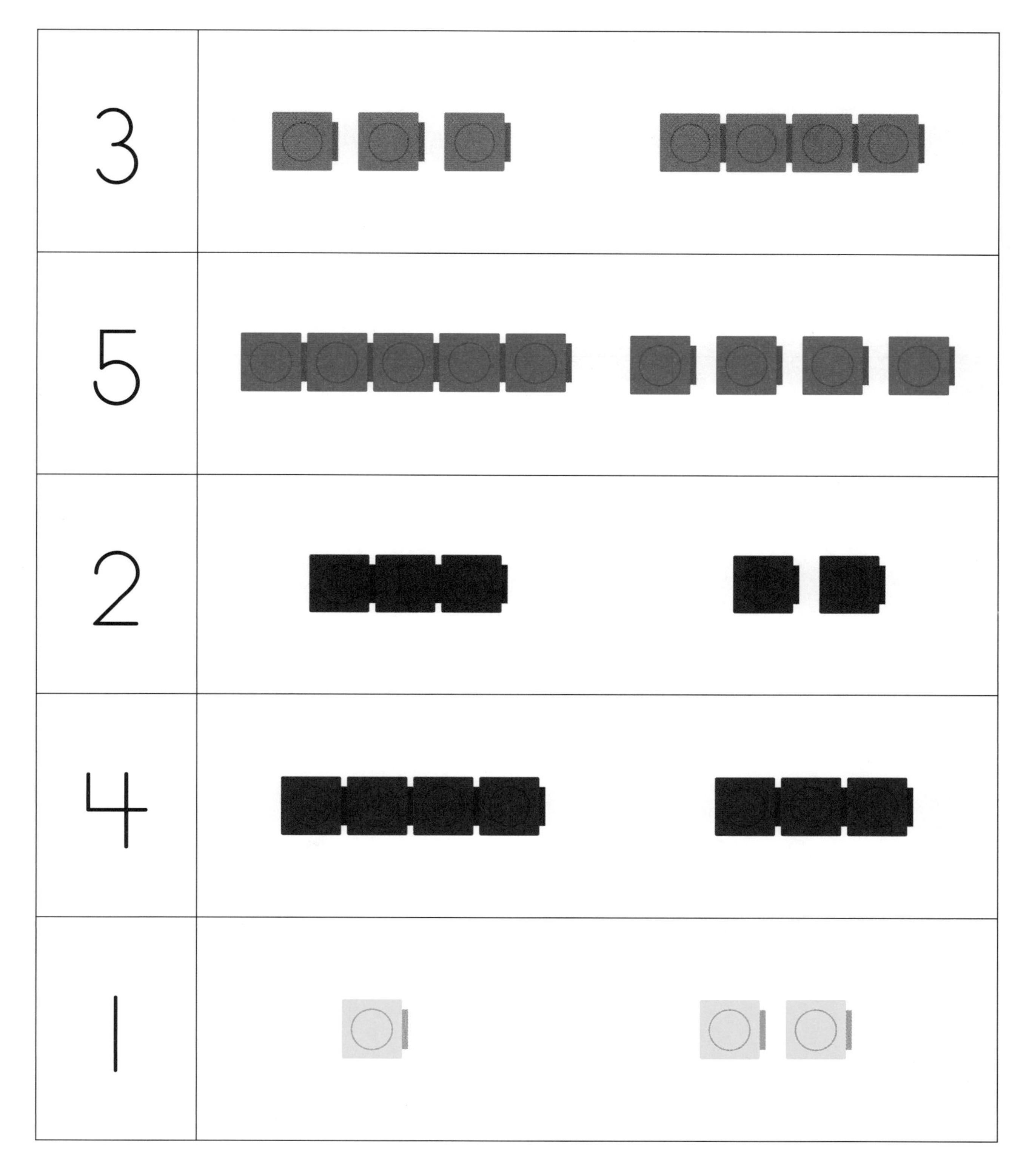

Using this page: Have students look at the numeral shown in each row and count each group of cubes, then circle the group that matches the number.
Concept: Numeral recognition 1 to 5 and cardinality.

Circle the correct number.

2 5 3	3 2 5
2 4 1	3 1 2
5 4 3	4 3 2

Using this page: Have students count the dots and circle the correct numeral.
Concept: Numeral recognition 1 to 5 and cardinality.

Color the correct number of things.

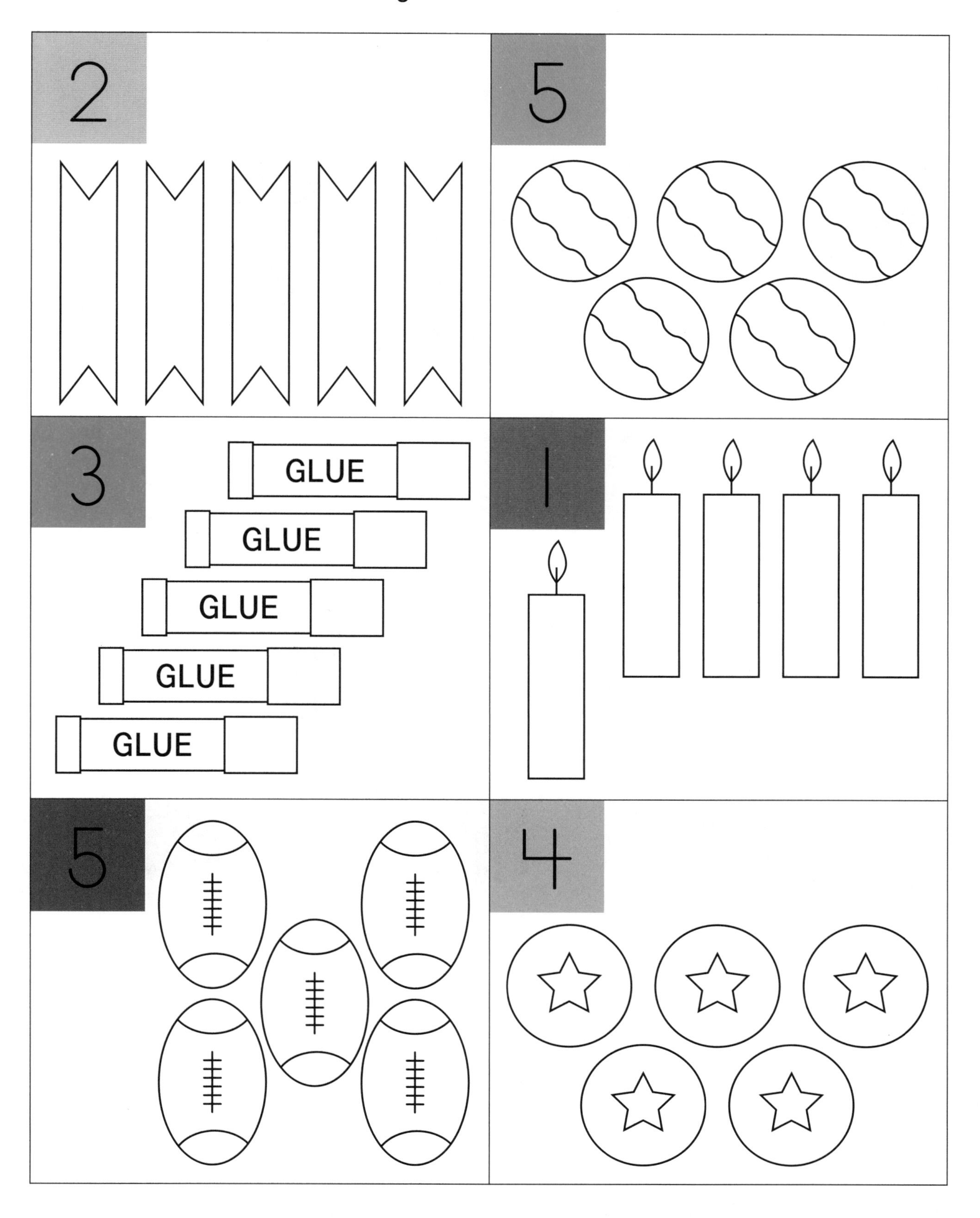

Using this page: Have students look at the numeral shown in each box and count that number of objects, then color them.
Concept: Numeral recognition 1 to 5 and cardinality.

Match.

Using this page: Have students count the stripes on each fish and match it to the character with the same numeral, then extend the fishing line from the character to the fish.
Concept: Numeral recognition 1 to 5 and cardinality.

Exercise 6

Trace and write 1.

Using this page: Have students count the umbrella, then trace the dotted lines on it and the raindrops. Next, have them trace and practice writing the numeral 1.
Concept: Numeral writing of 1.

Trace and write 2.

Using this page: Have students count, then trace the dotted lines around the swans. Next, have them trace and practice writing the numeral 2.
Concept: Numeral writing of 2.

Exercise 7

Trace and write 3.

Using this page: Have students count, then trace the dotted lines around the snowmen. Next, have them trace and practice writing the numeral 3.
Concept: Numeral writing of 3.

Count and write the number.

Using this page: Have students count the objects and write the numeral in the box.
Concept: Numeral writing of 1 to 3 and cardinality.

Exercise 8

Trace and write 4.

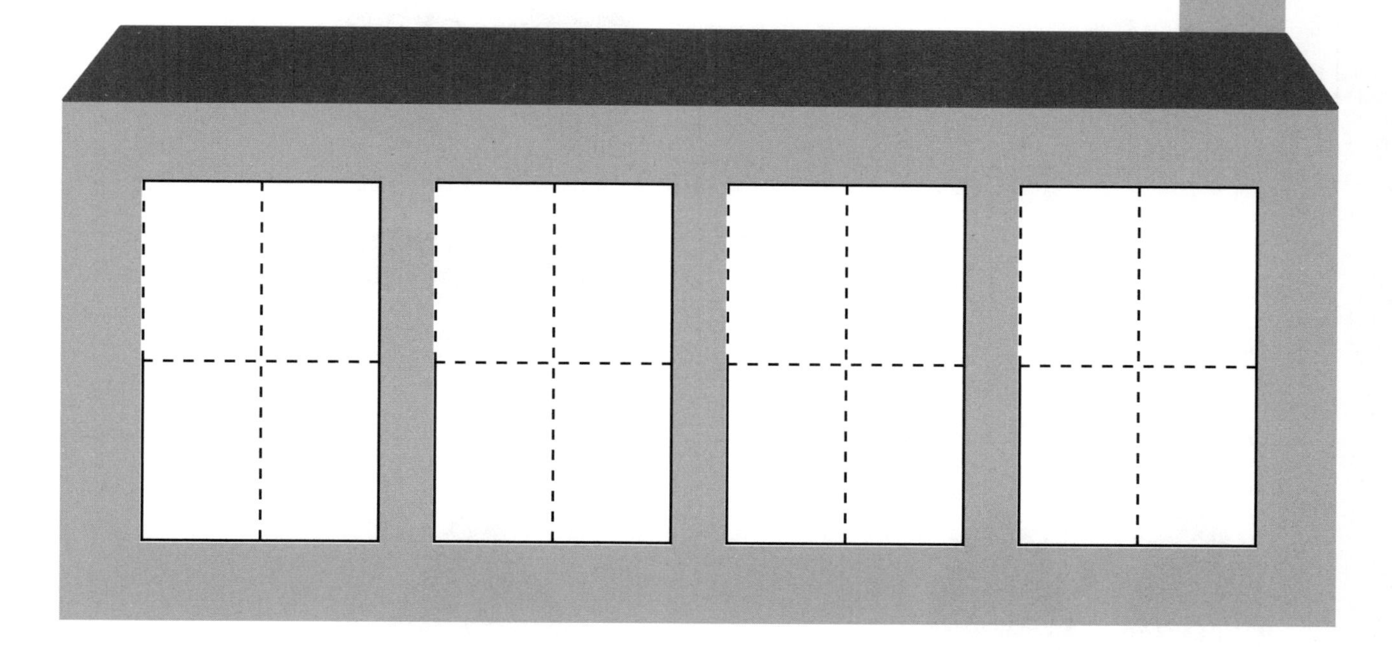

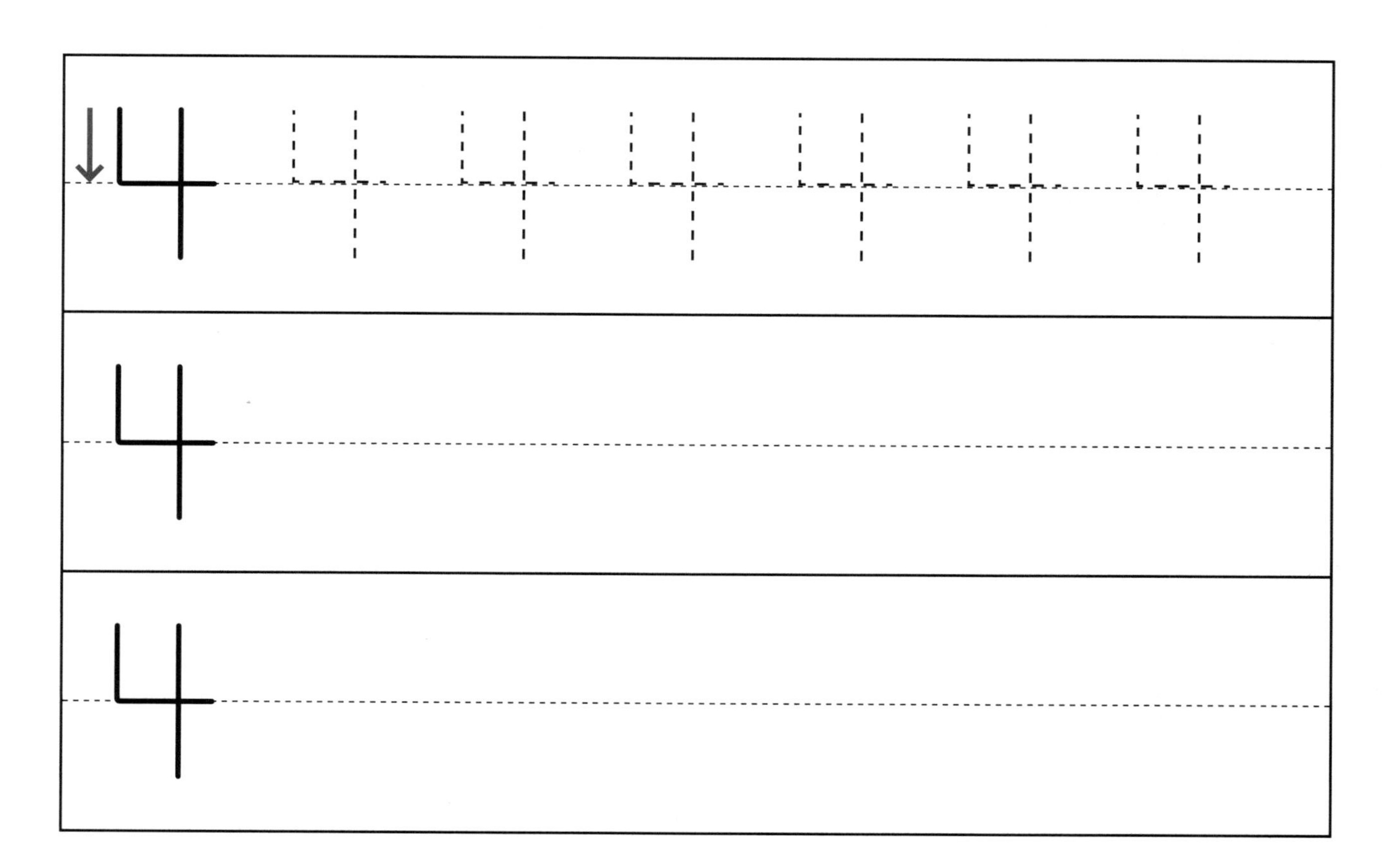

Using this page: Have students count, then trace the dotted lines around the windows. Next, have them trace and practice writing the numeral 4.
Concept: Numeral writing of 4.

Count and write the number.

Using this page: Have students count the objects and write the numeral in the box.
Concept: Numeral writing of 1 to 4 and cardinality.

Exercise 9

Trace and write 5.

Using this page: Have students count, then trace the dotted lines around the seahorses. Next, have them trace and practice writing the numeral 5.
Concept: Numeral writing of 5.

Count and write the number.

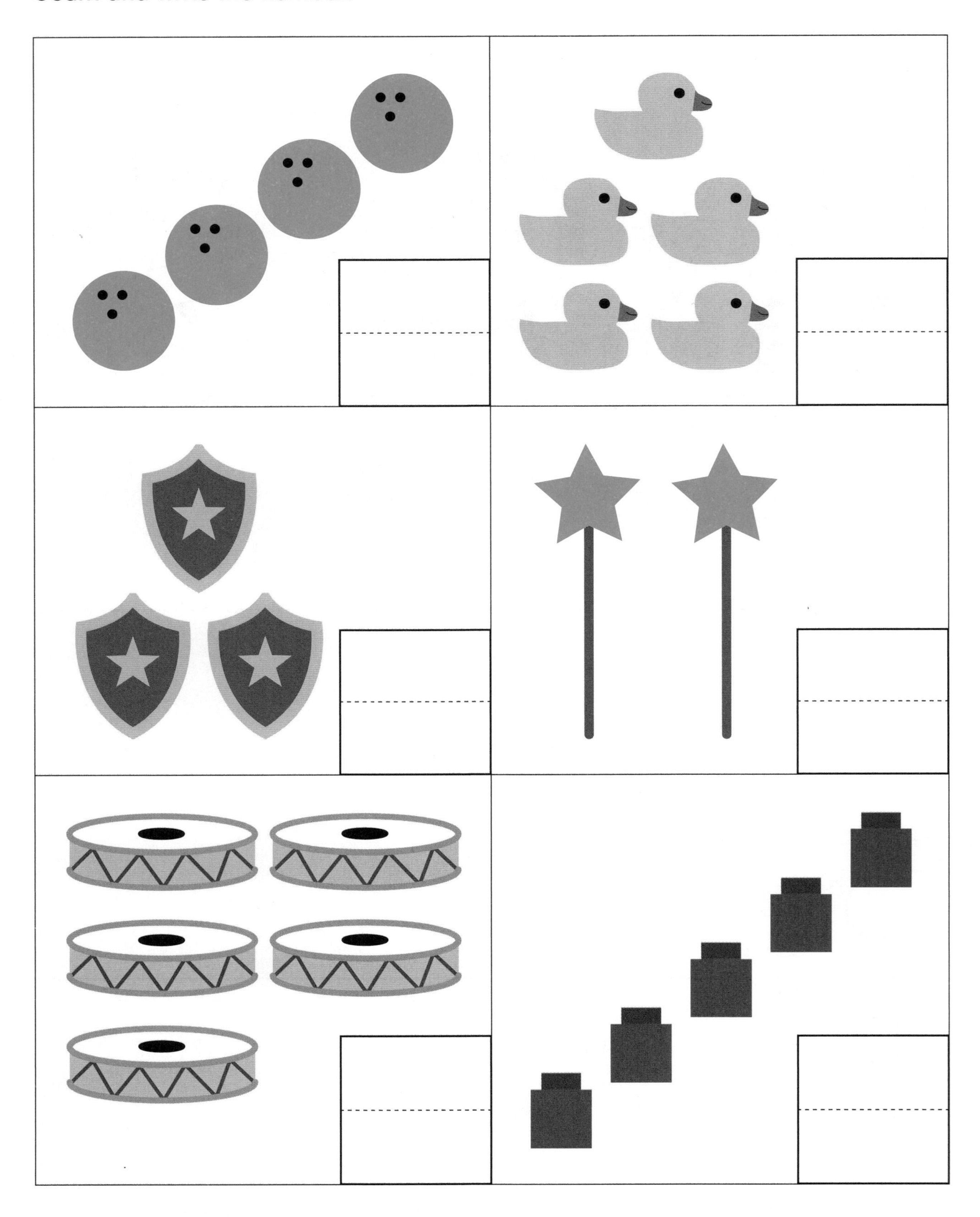

Using this page: Have students count the objects and write the numeral in the box.
Concept: Numeral writing of 1 to 5 and cardinality.

Exercise 10

Trace and write 0.

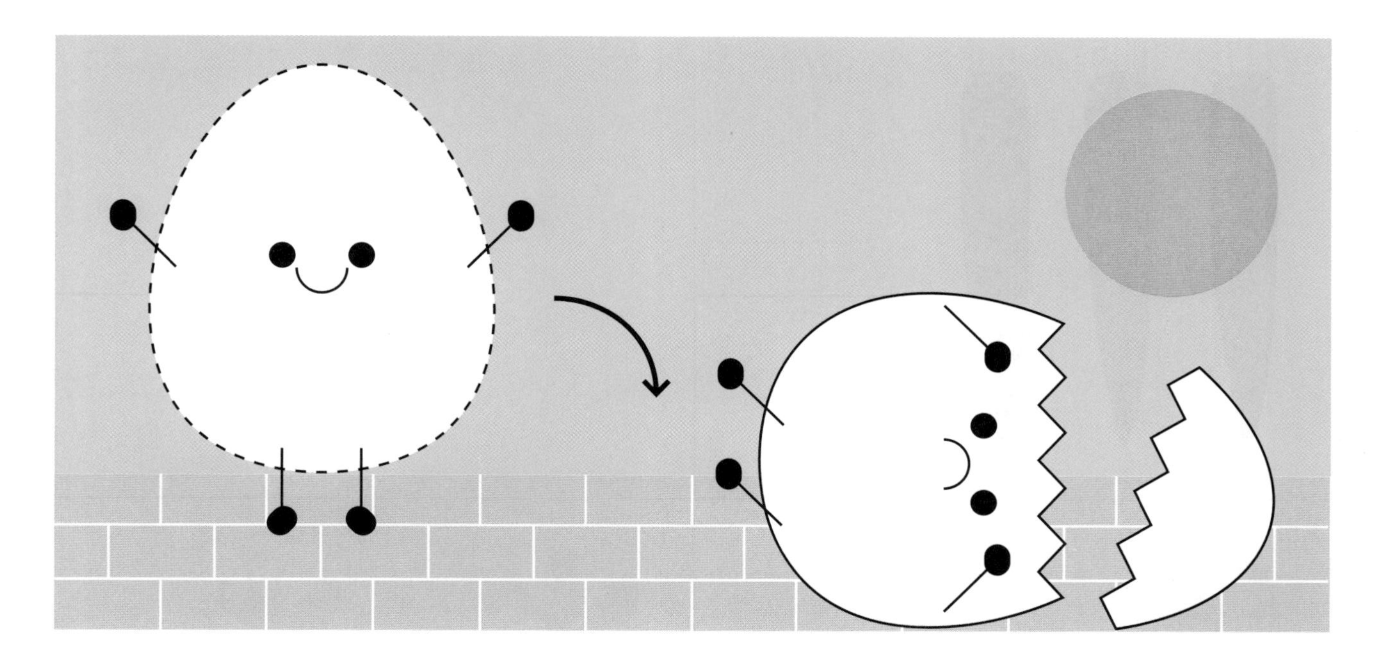

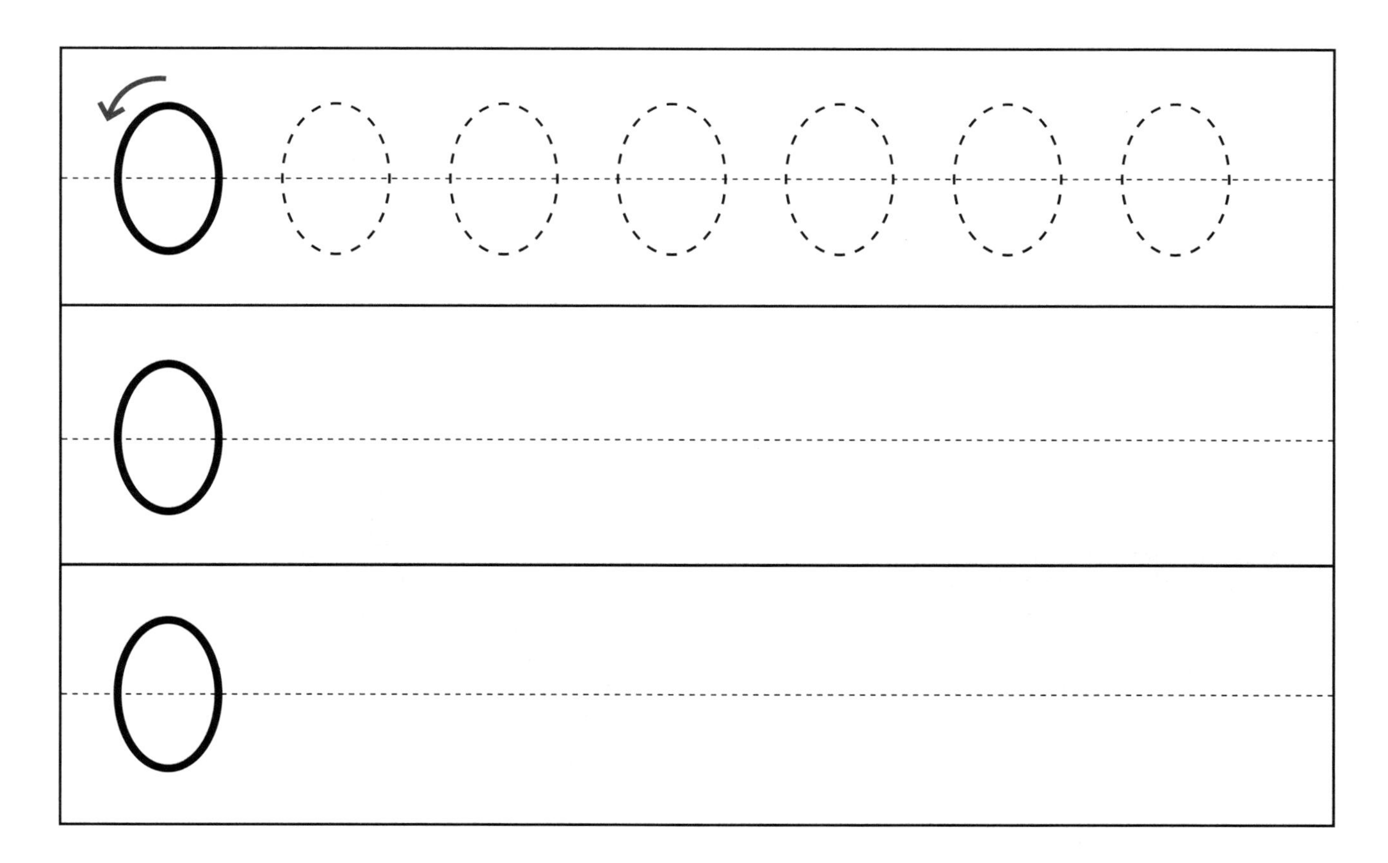

Using this page: Have students trace the egg, then determine how many eggs are left after it fell and broke. Next, have students trace and practice writing the numeral 0.
Concept: Numeral writing of 0.

Count and write the number.

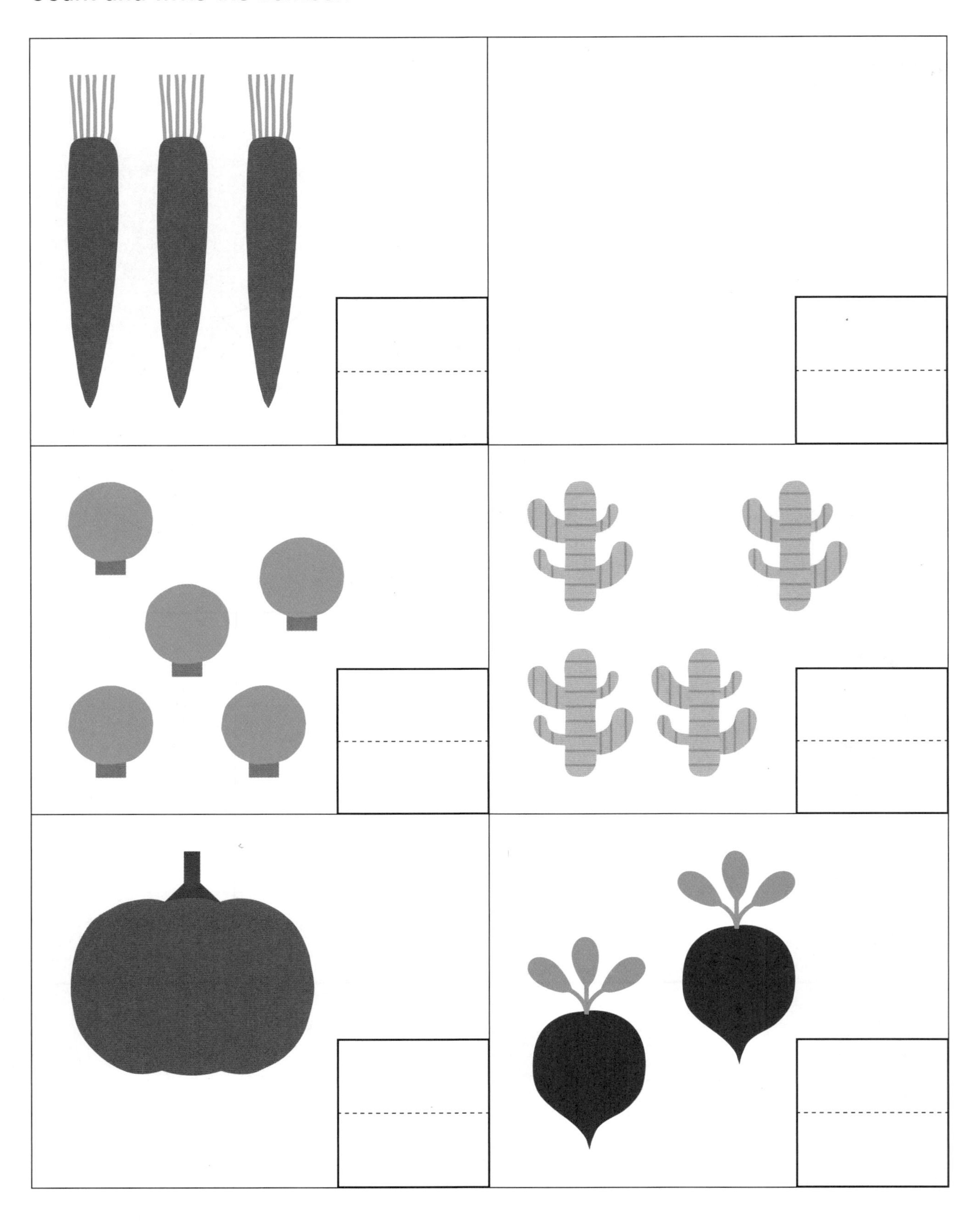

Using this page: Have students count the objects and write the numeral in the box.
Concept: Numeral writing of 0 to 5.

Exercise 11

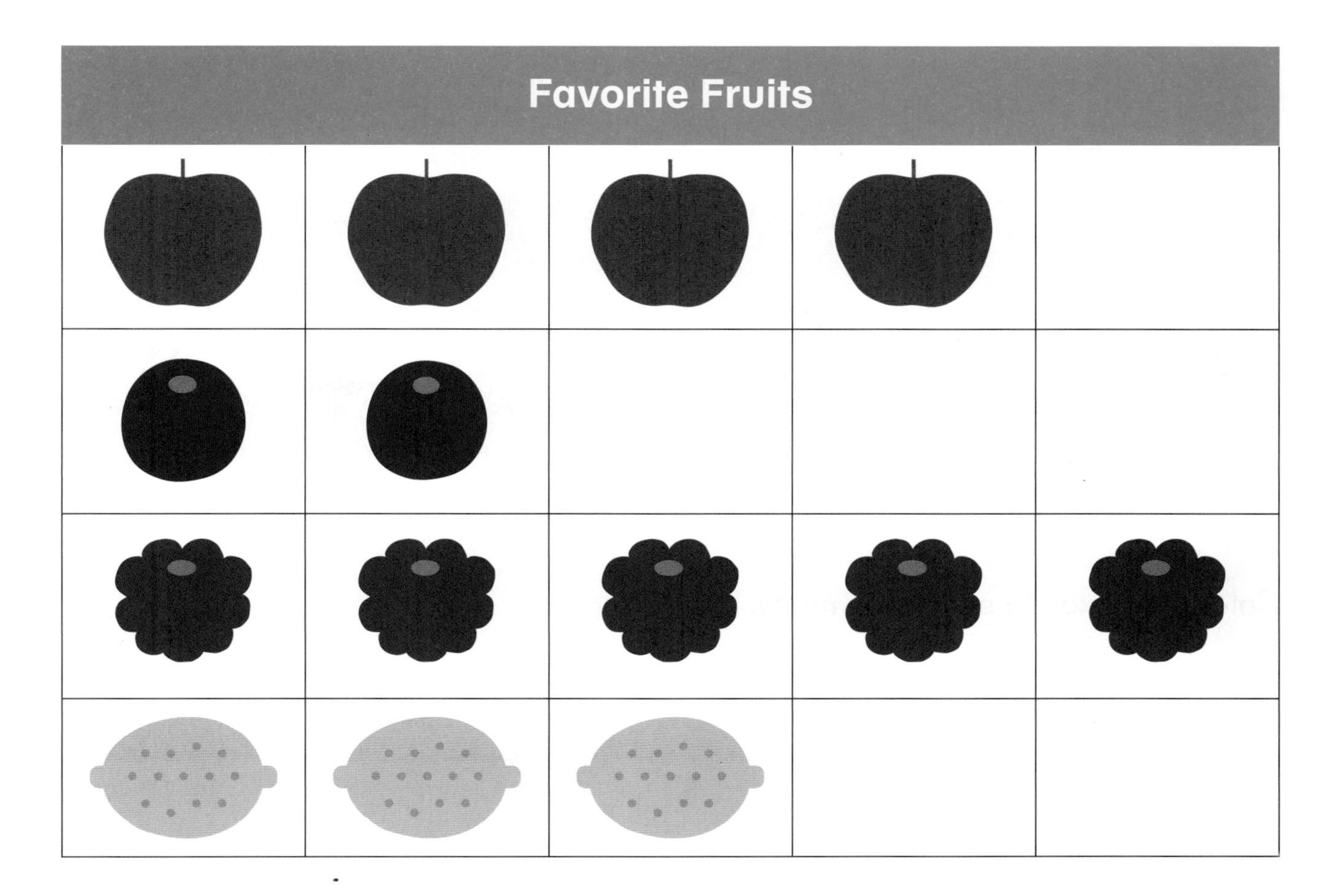

Write the number to show how many.

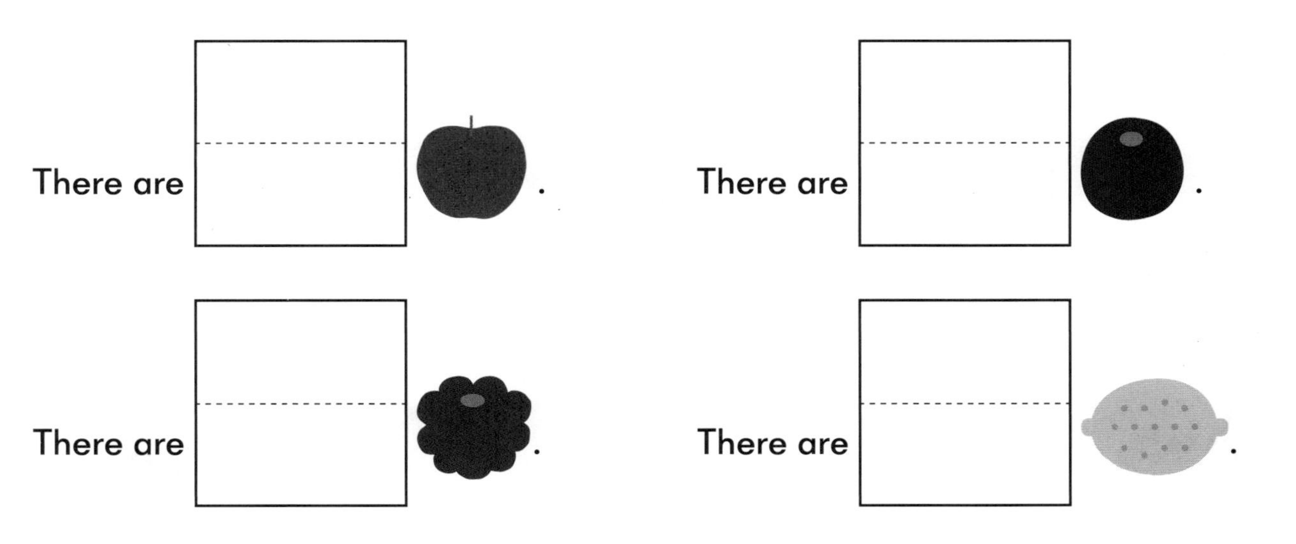

Using this page: Have students count the fruit in the picture graph and write the numeral in the box.
Concept: Reading picture graphs and numeral writing 1 to 5.

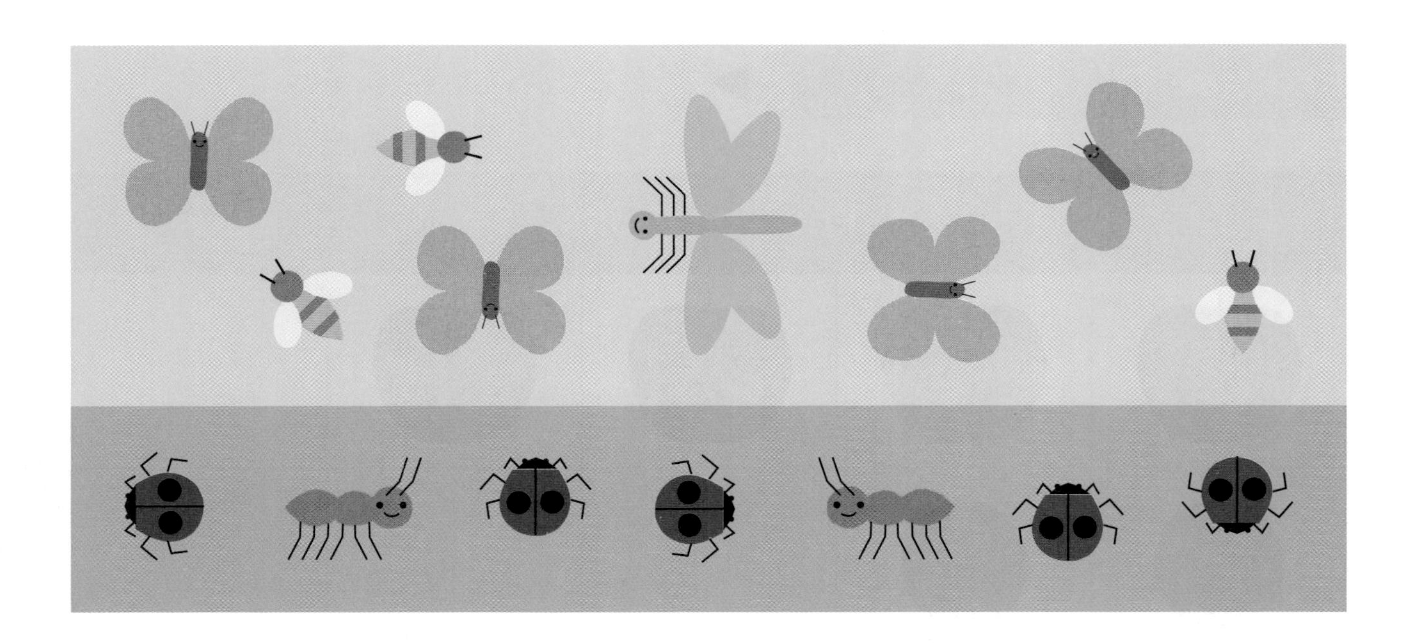

Color the boxes to show how many.

Bugs in the Garden				

Using this page: Have students cross out the bugs as they count and color that number of boxes.
Concept: Representing data on a graph.

Exercise 12

Match.

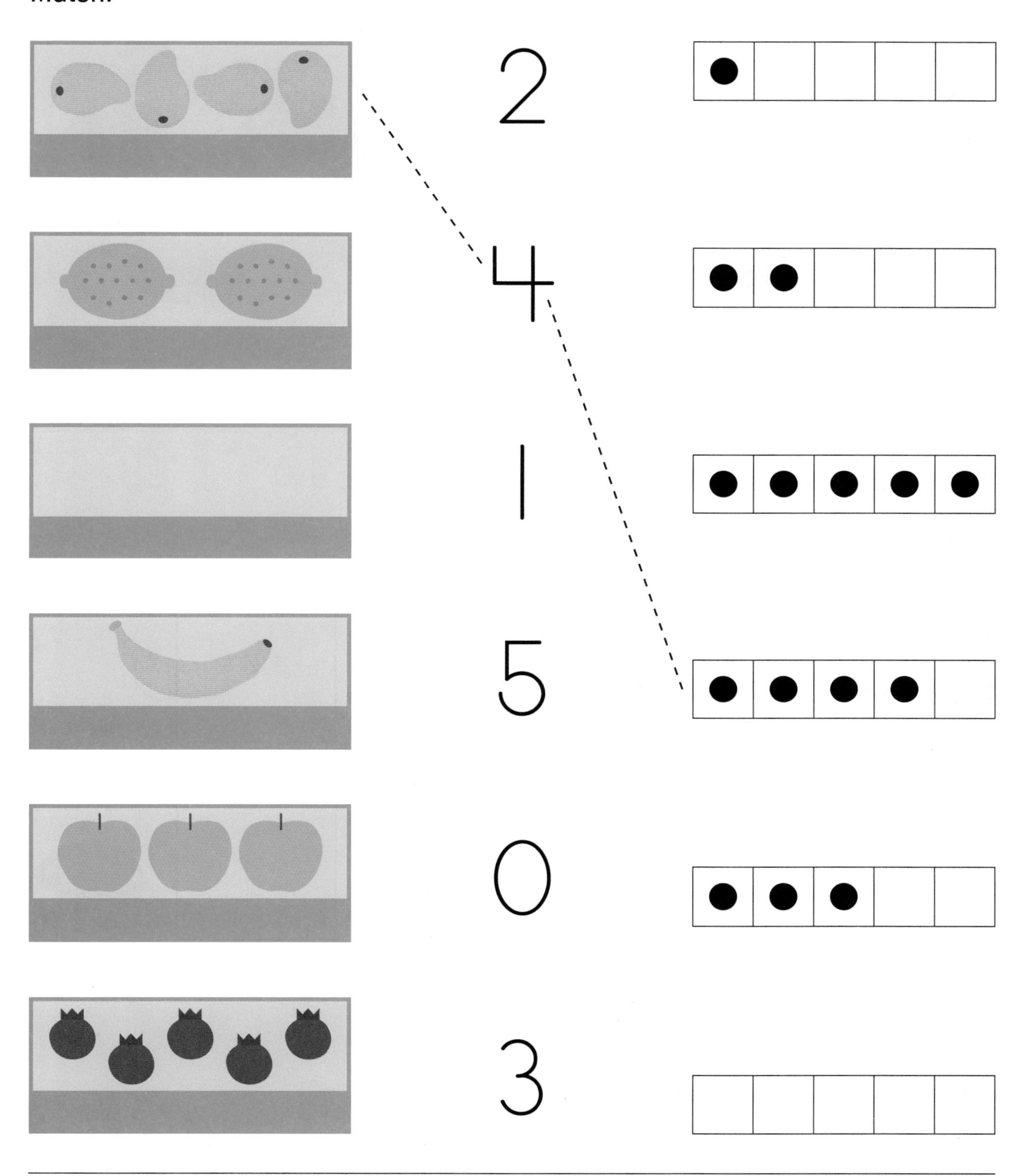

Using this page: Have students match the objects and five-frame cards to the numerals.

Write the number to show how many.

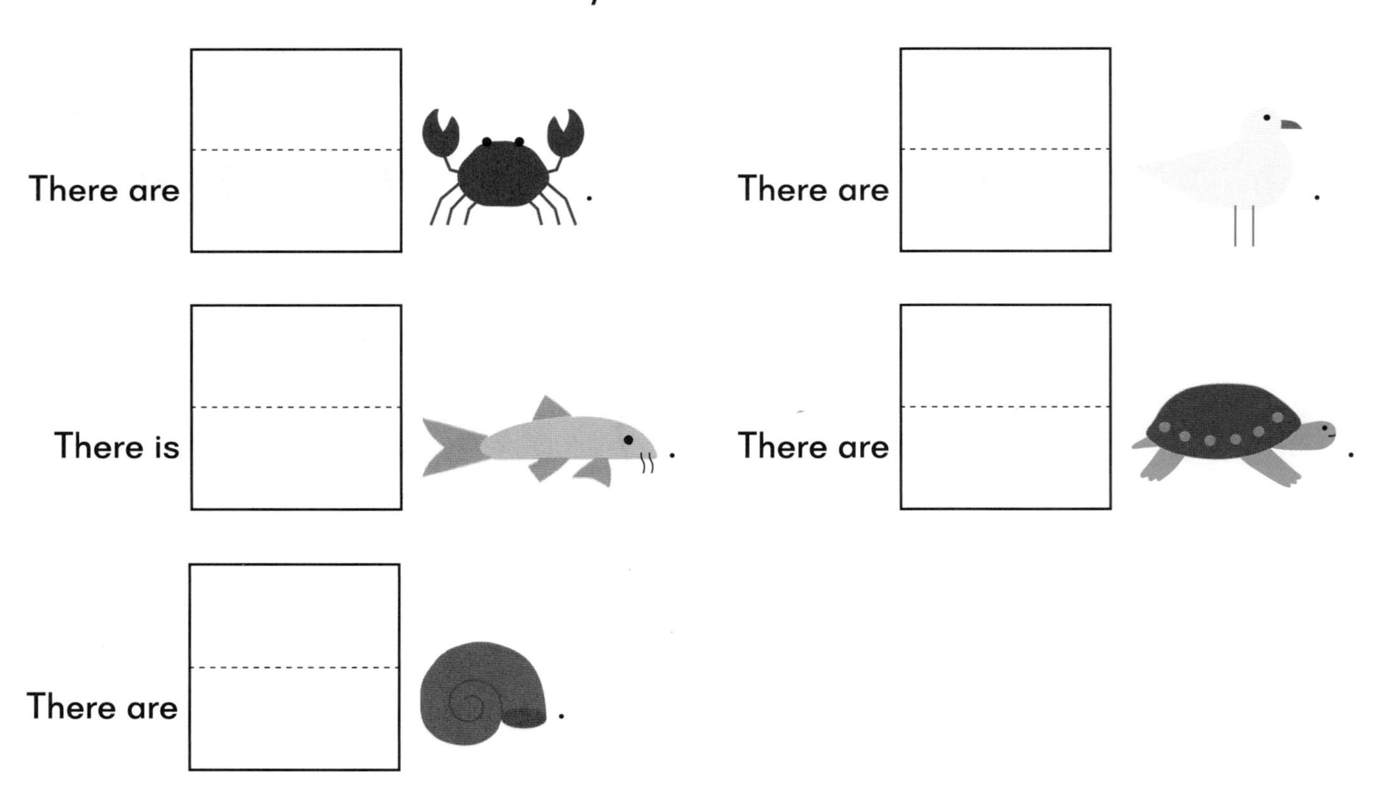

Using this page: Have students count the animals and write the numeral in the box.

Color in the picture graph.

Vegetables in the Garden				

Using this page: Have students count one type of vegetable at a time and color the boxes in that column to show the number.

Color according to the Color Key.

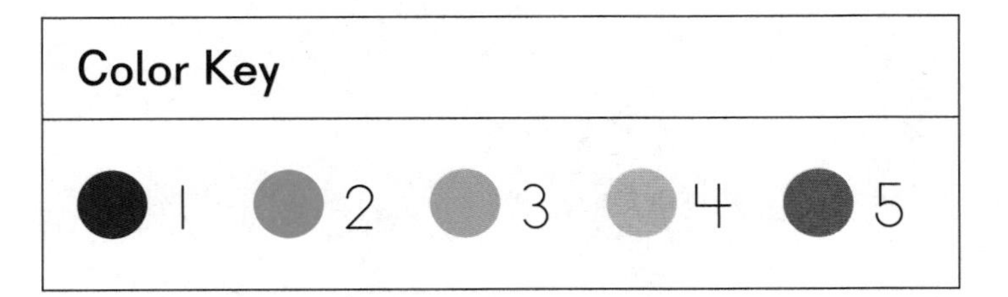

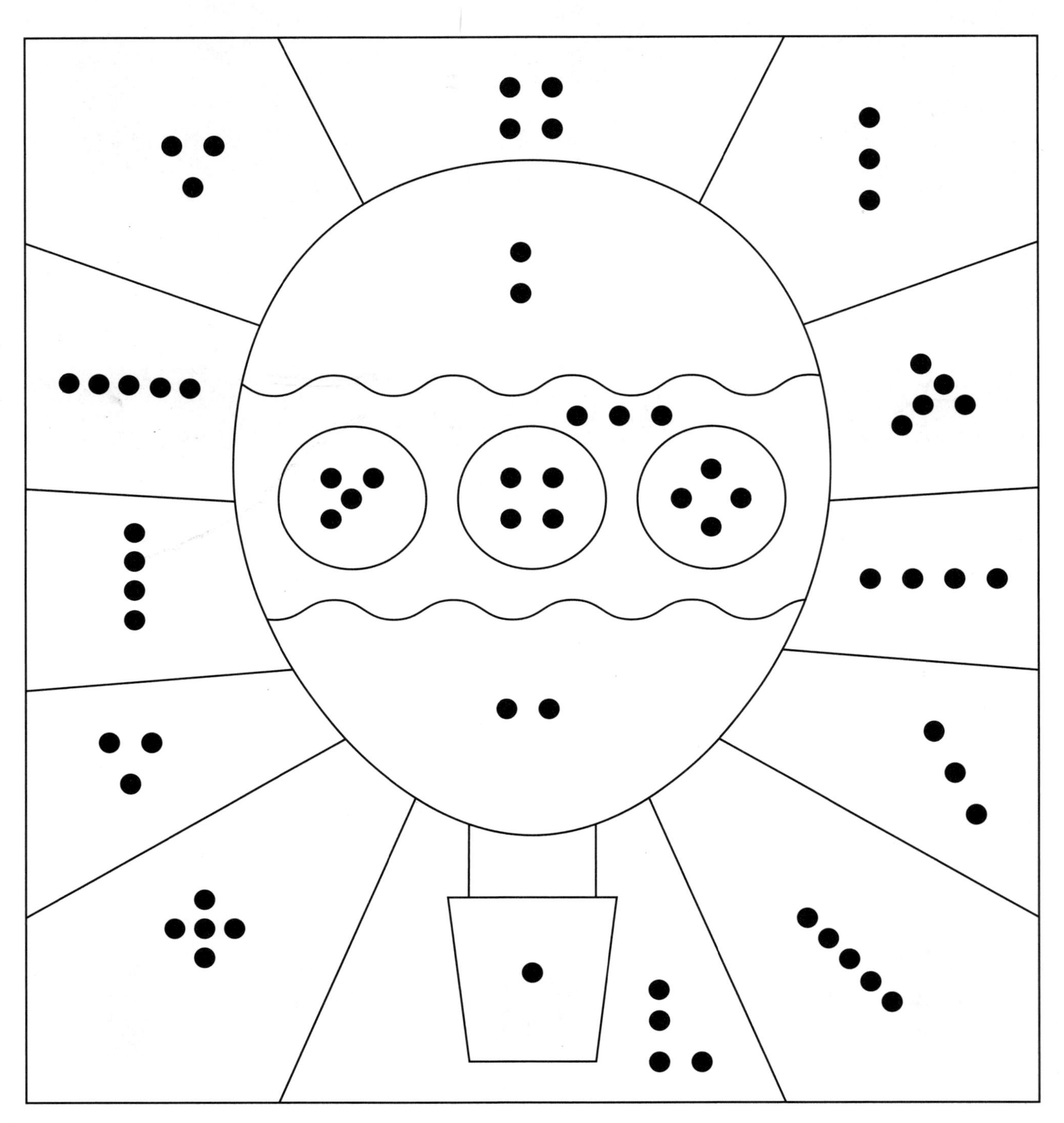

Using this page: Have students count the dots and color them with the color specified in the color key.

Chapter 3 Numbers to 10

Exercise 1

Circle the groups of 6.

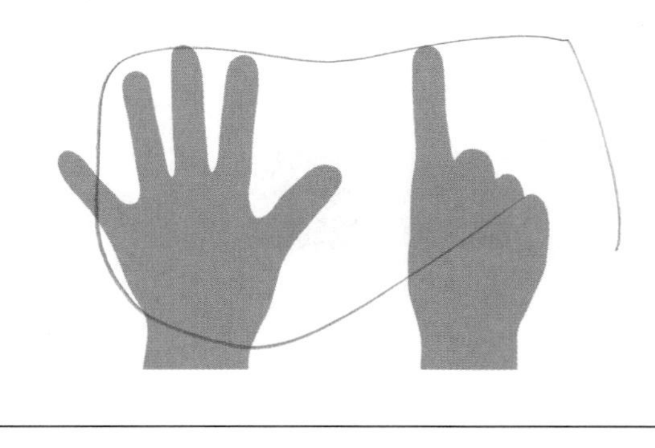

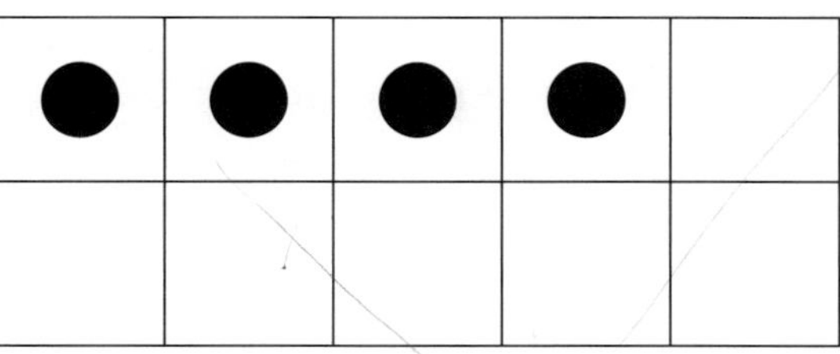

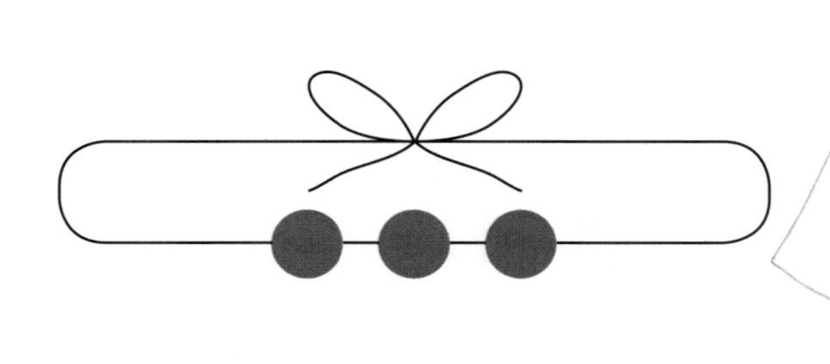

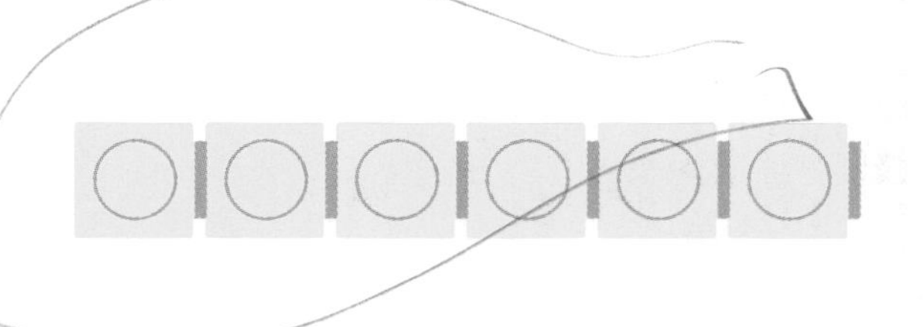

Using this page: Have students count and circle the groups of objects that have six.
Concept: Identifying sets of six.

Write an X in the boxes with 7 things.

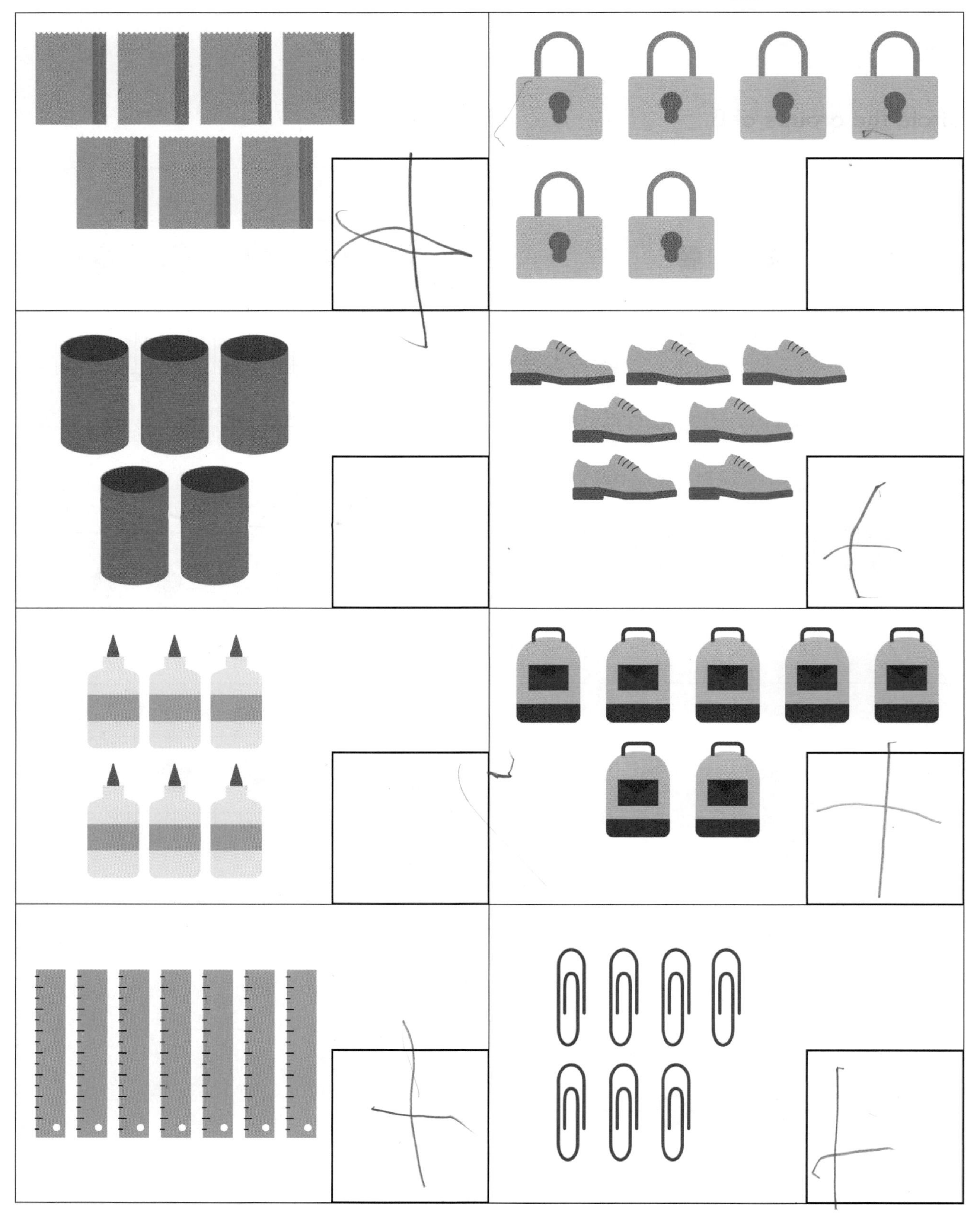

Using this page: Have students cross out the objects as they count, then write an X in each box that has seven objects.
Concept: Identifying sets of seven.

Exercise 2

Circle the groups of 8.

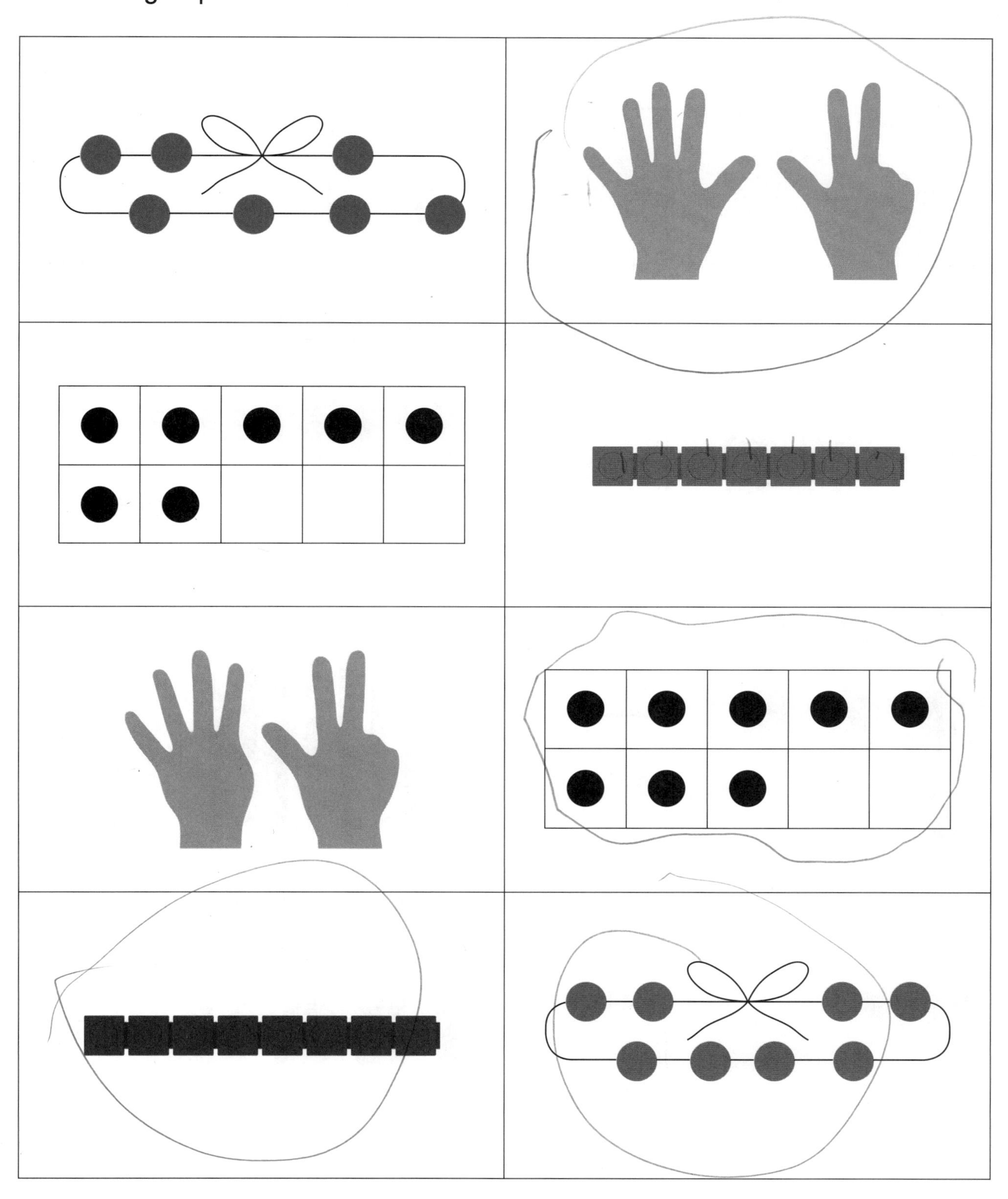

Using this page: Have students count and circle the groups of objects that have eight.
Concept: Identifying sets of eight.

Write an X in the boxes with 9 things.

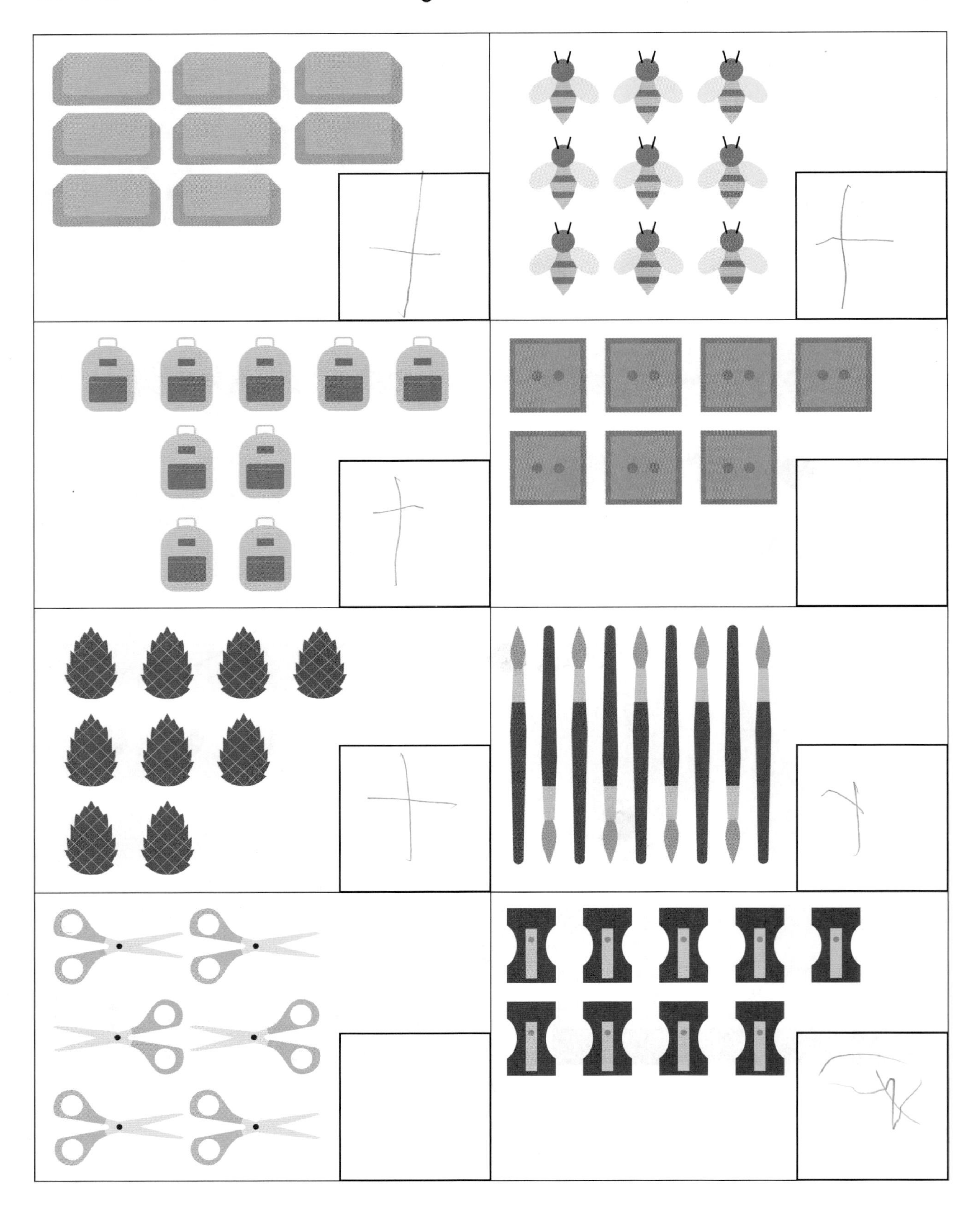

Using this page: Have students cross out the objects as they count, then write an X in each box that has nine objects.
Concept: Identifying sets of nine.

Exercise 3

Match.

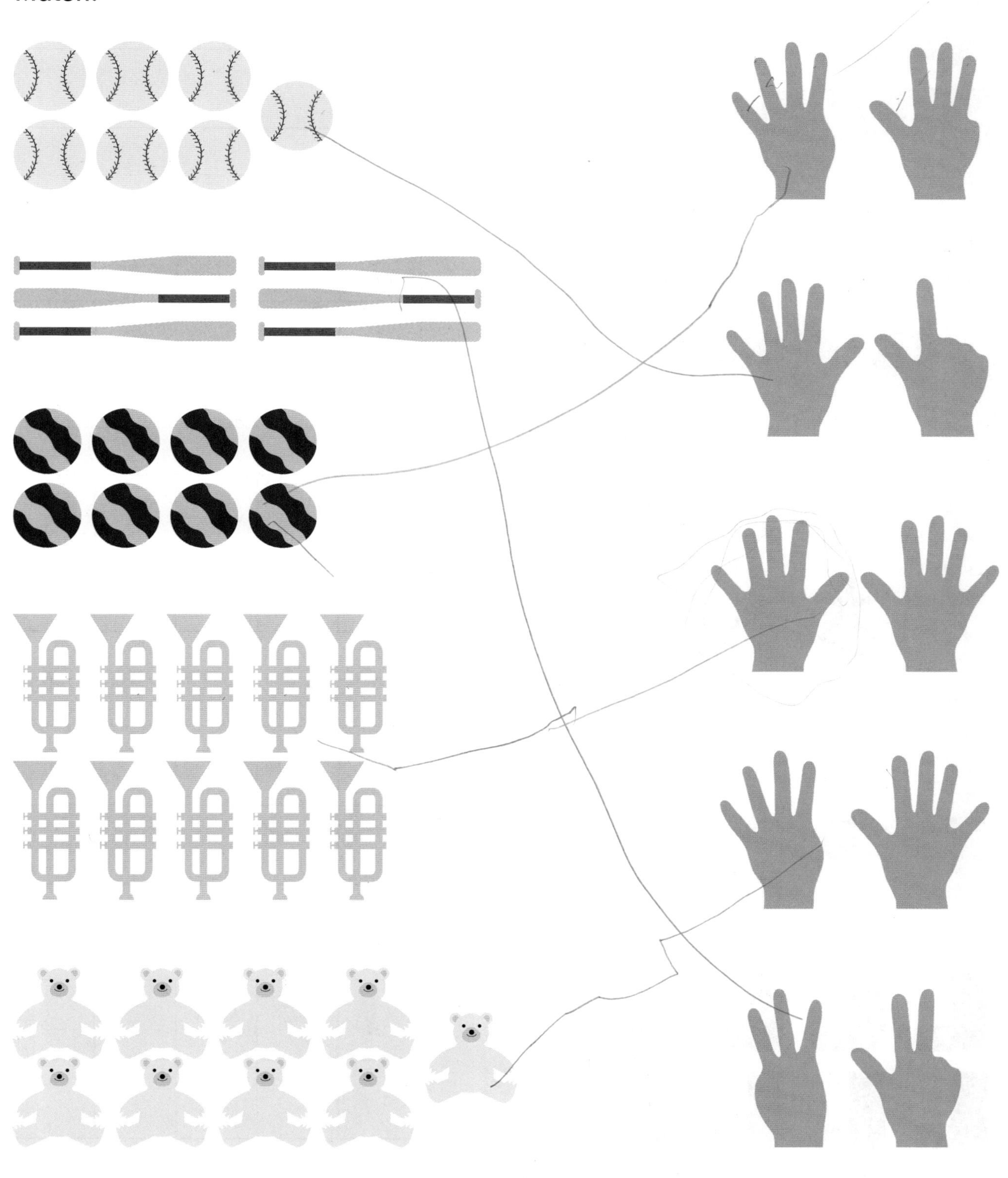

Using this page: Have students cross out the objects as they count, then match with the fingers showing that number.
Concept: Identifying sets of up to 10.

Count and color the correct number of boxes.

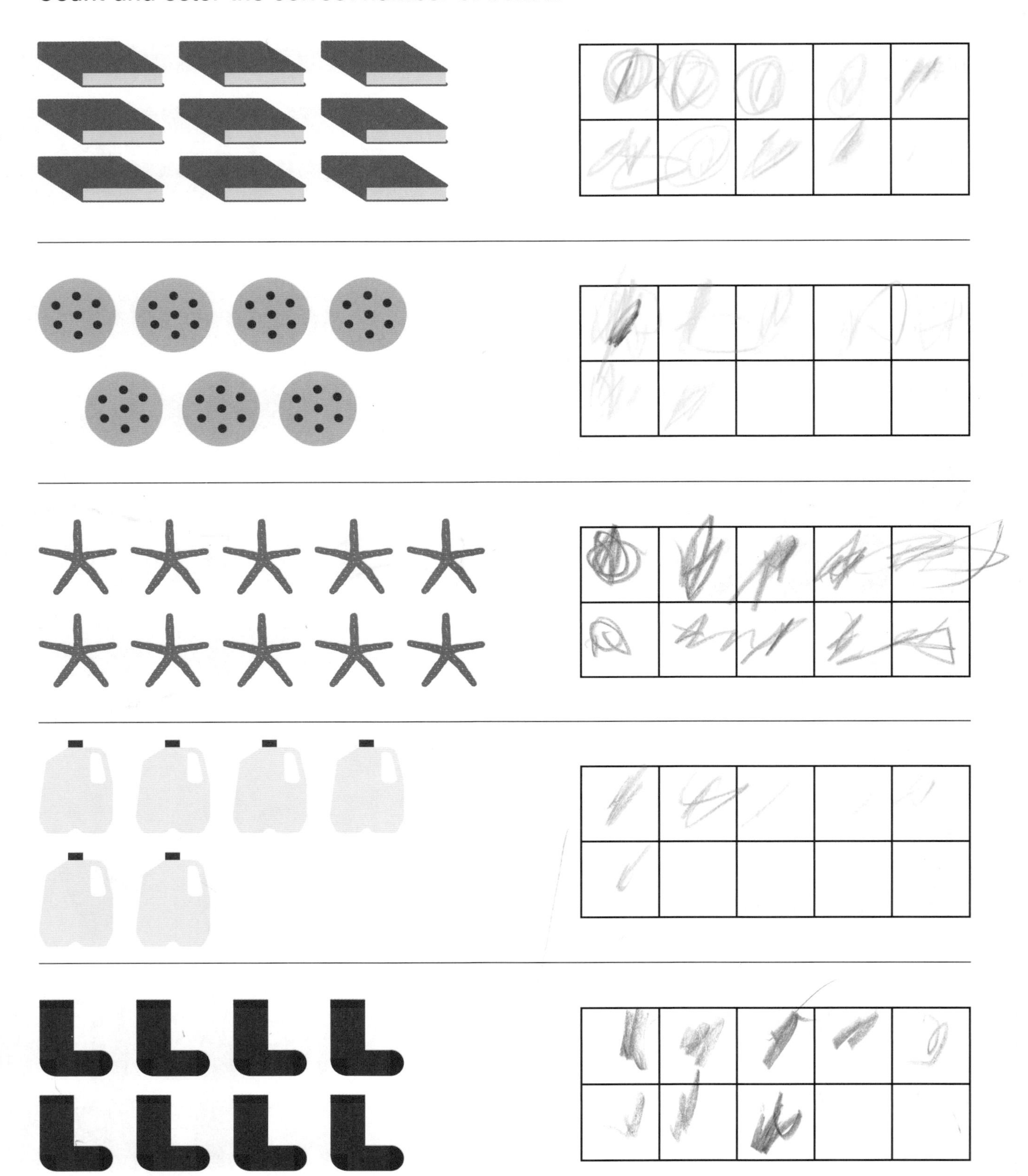

Using this page: Have students cross out the objects as they count, then color the boxes to show that number.
Concept: Identifying sets of up to 10.

Exercise 4

Circle the groups with the same number.

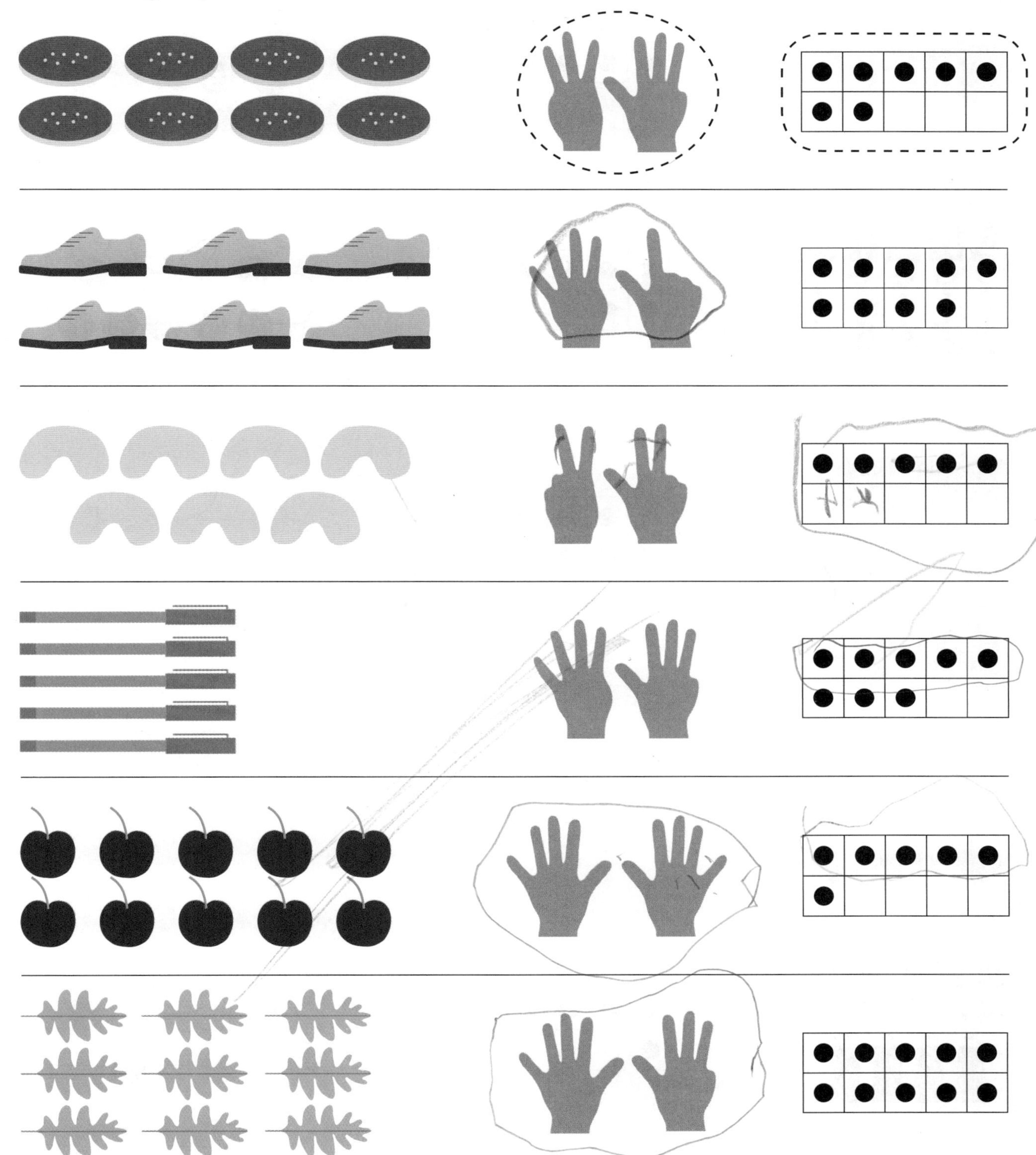

Using this page: Have students compare the number of objects, fingers, and dots on the ten-frames, then circle the groups with the same number.
Concept: Identifying sets of up to 10.

Match.

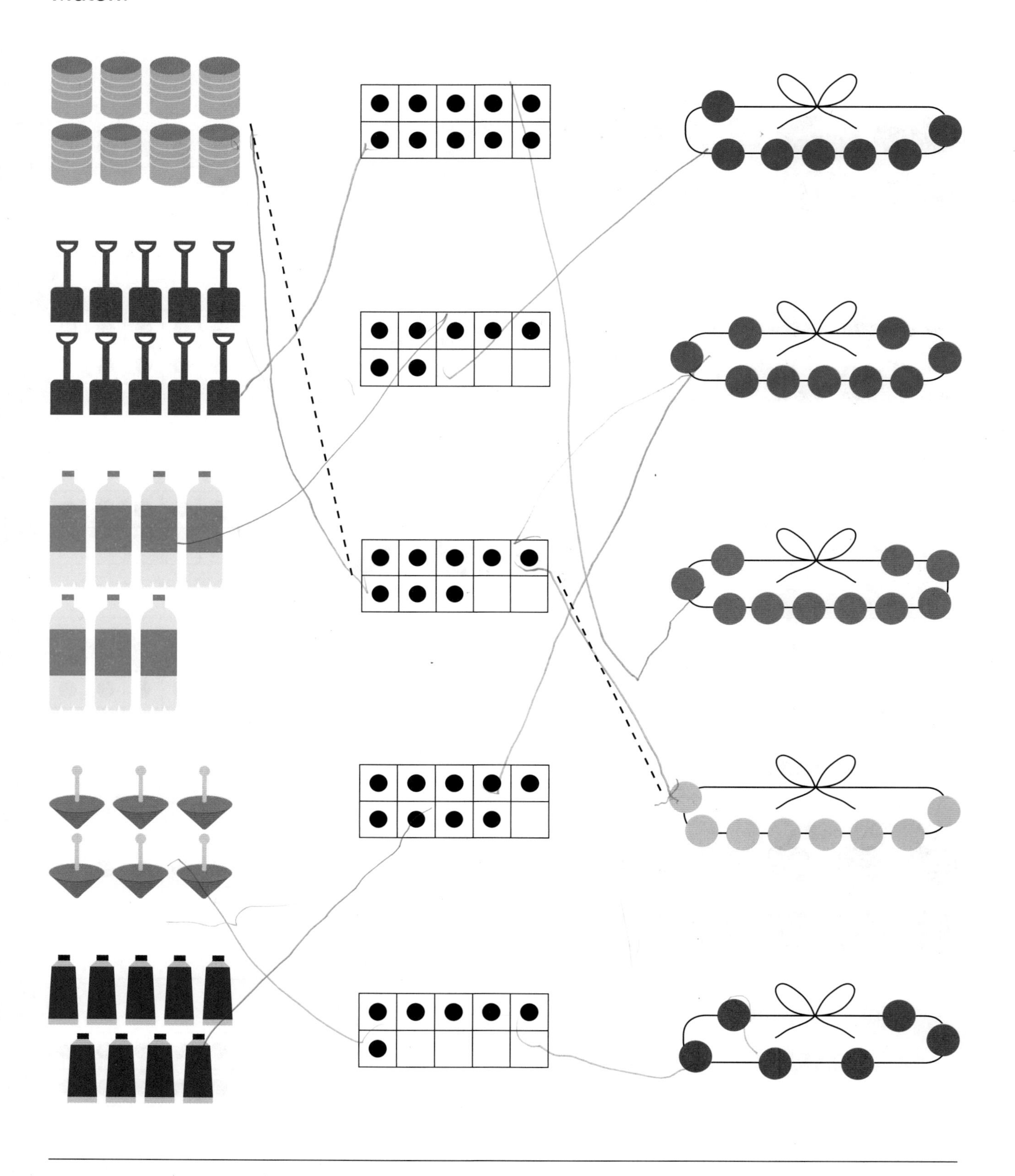

Using this page: Have students count the objects and match with the ten-frame card, then count and match the beads to the ten-frame card.
Concept: Identifying sets of up to 10.

Connect the dots.

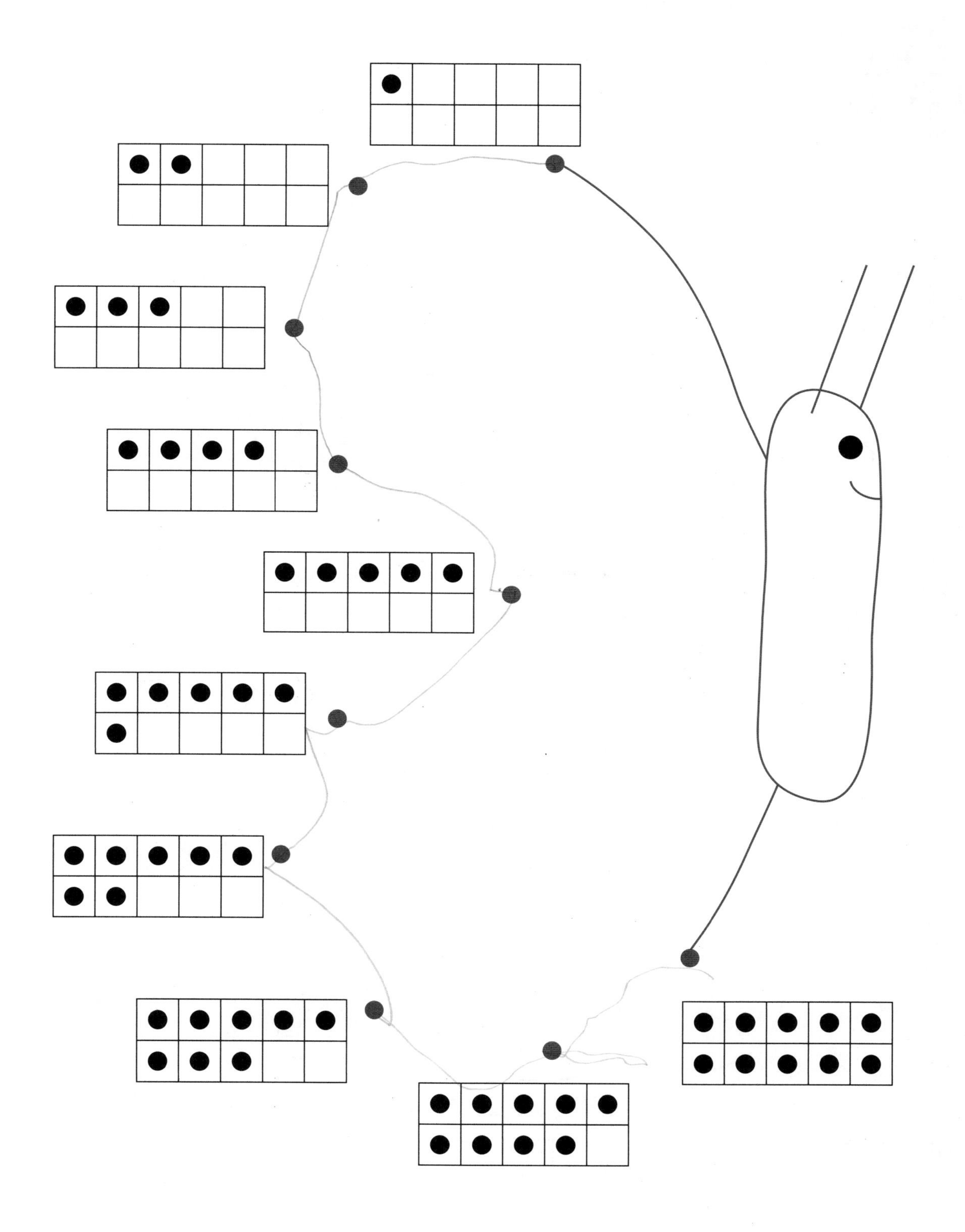

Using this page: Have students count the dots on the ten-frame cards and connect them to complete the picture.
Concept: Identifying sets of up to 10.

Find the 10 balls and color.

Using this page: Have students find the ten balls and color.
Concept: Counting to 10.

Exercise 5

Circle the correct number.

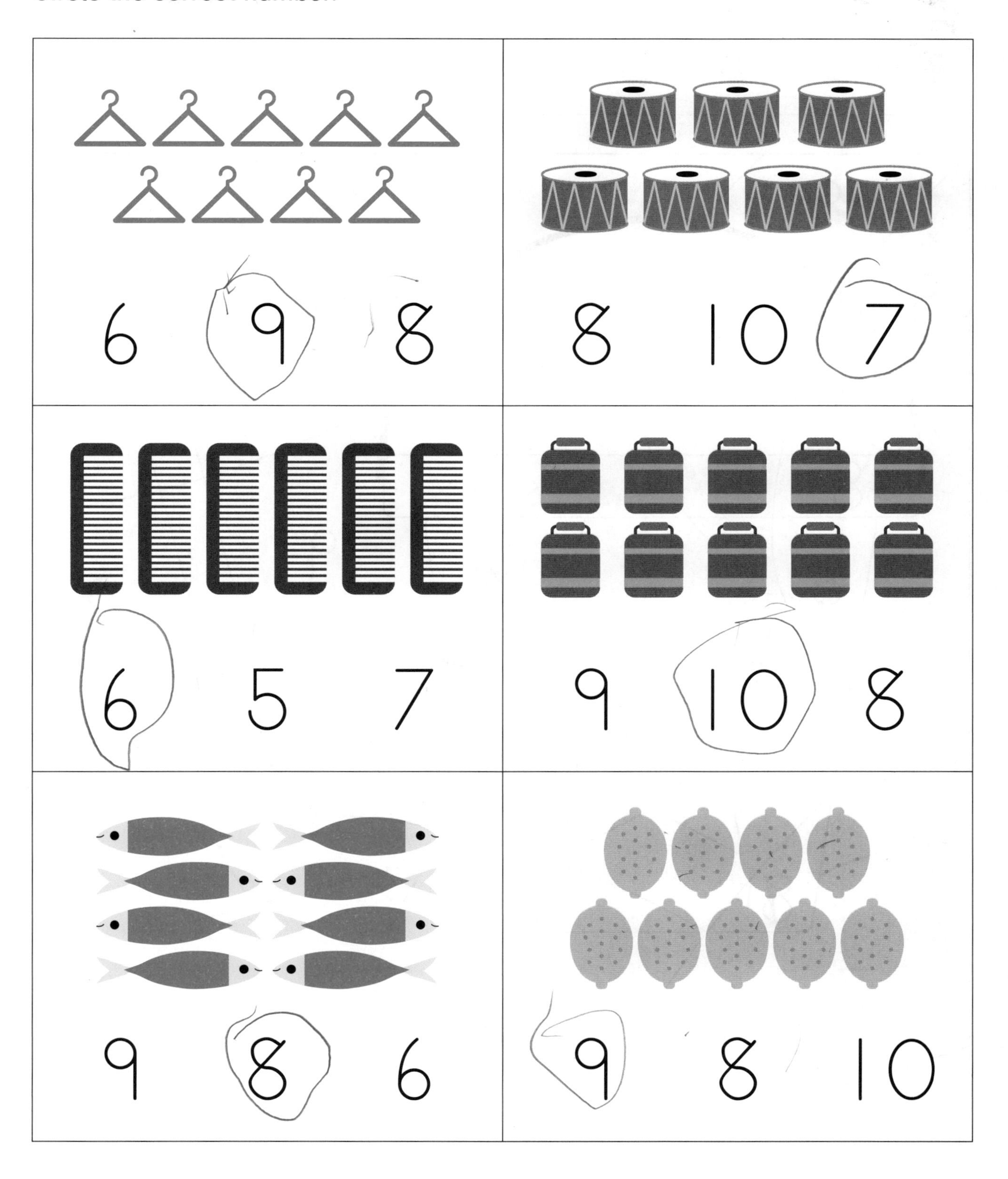

Using this page: Have students count the objects, then circle the correct numeral.
Concept: Numeral recognition 6 to 10.

Color according to the Color Key.

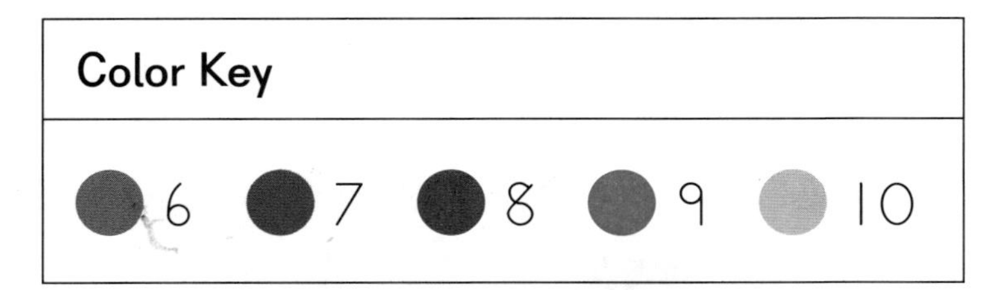

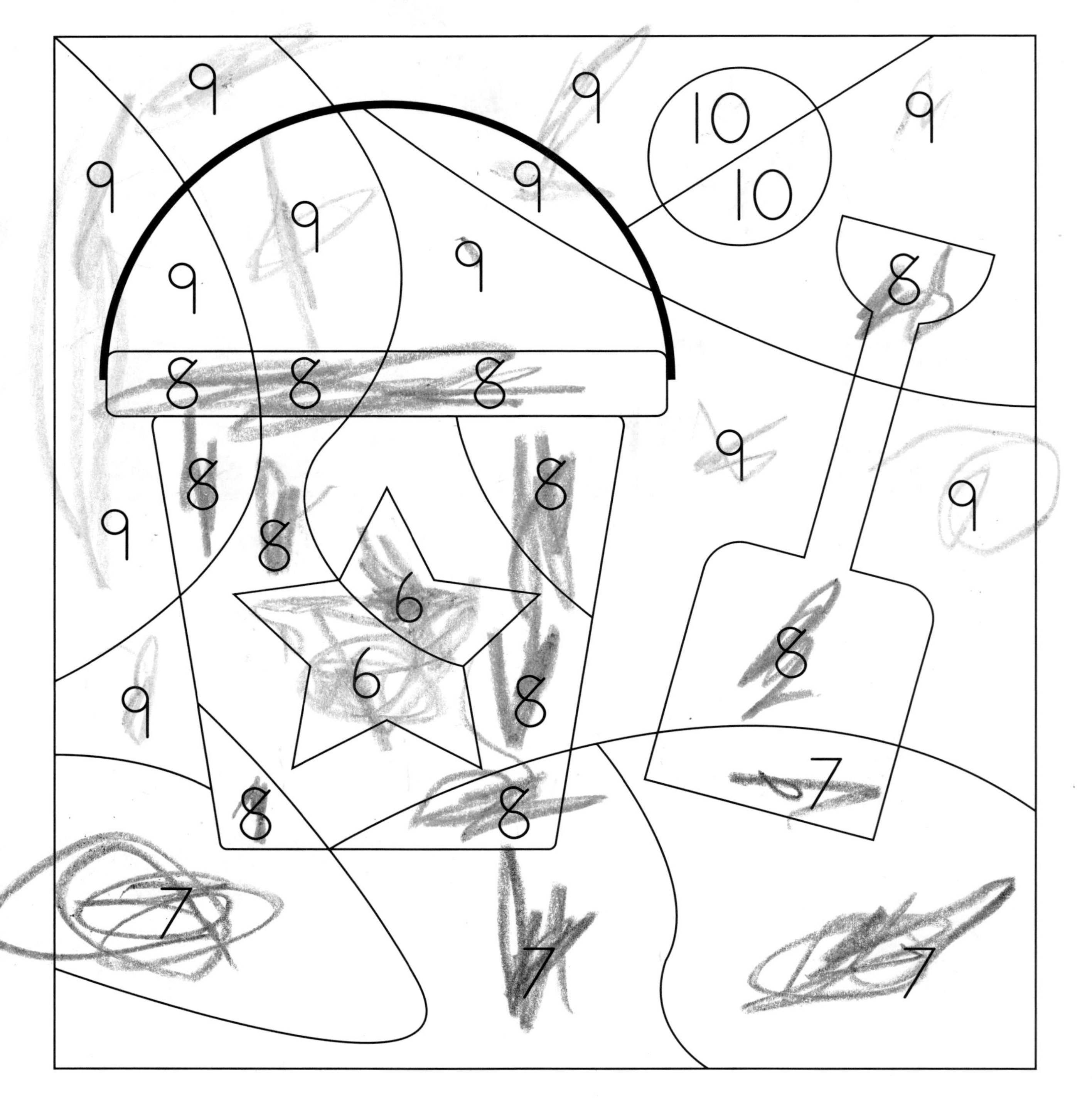

Using this page: Have students refer to the color key and color the picture.
Concept: Numeral recognition 6 to 10.

Exercise 6

Trace and write 6.

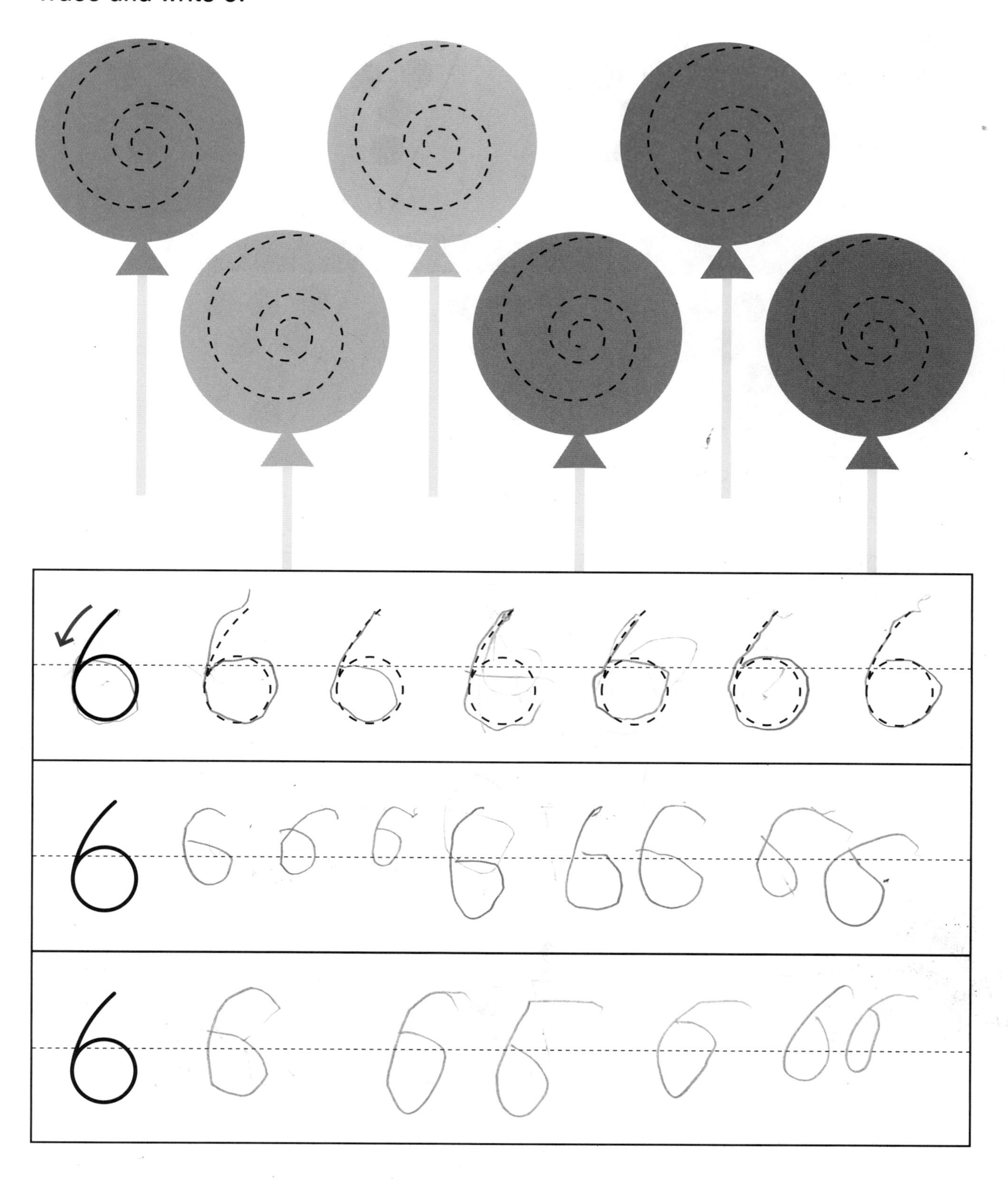

Using this page: Have students count the lollipops and trace the spirals. Then have them trace and practice writing numeral 6.
Concept: Writing of numeral 6.

Trace and write 7.

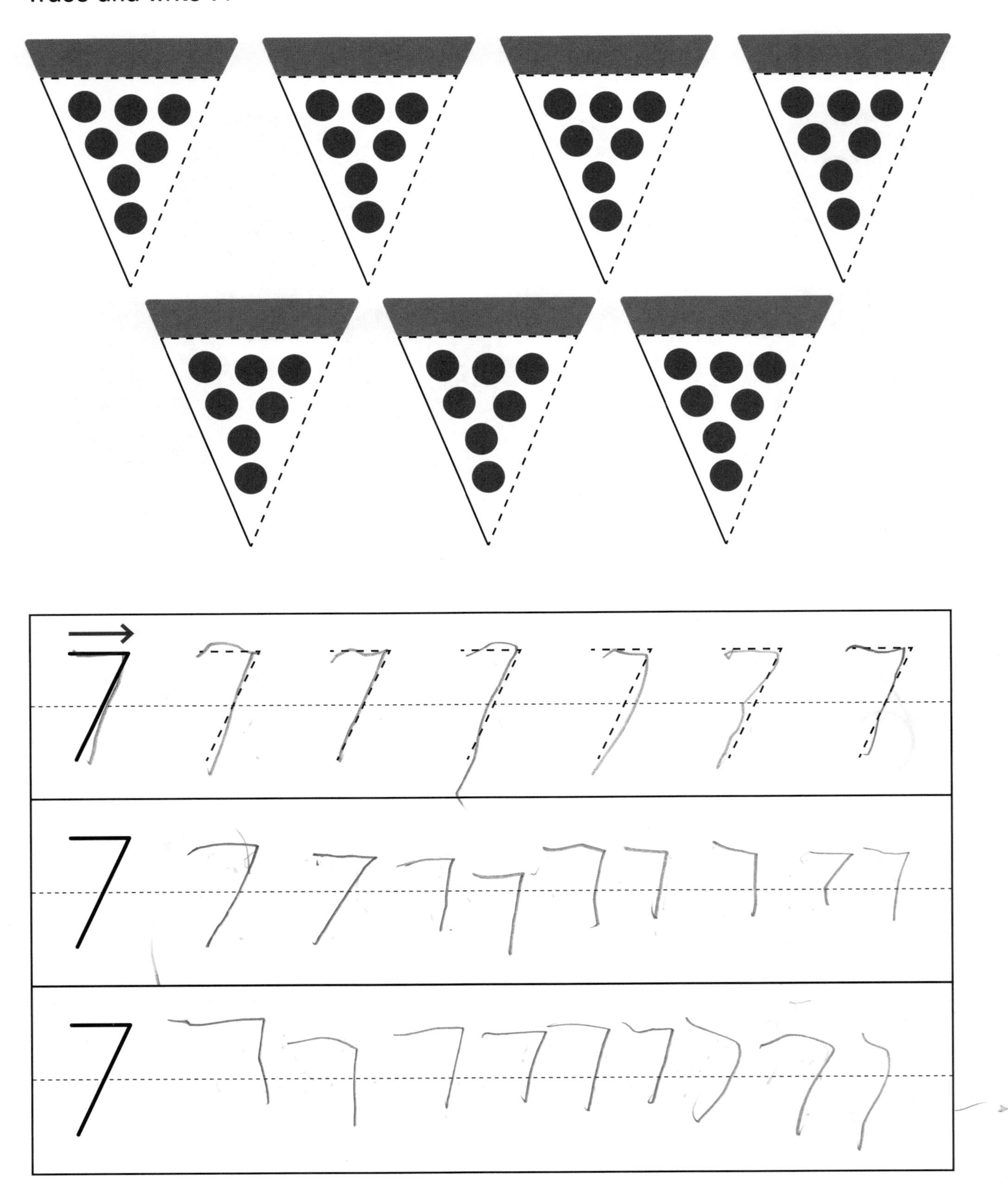

Using this page: Have students count and trace the dotted lines around the slices of pizza. Then have them trace and practice writing numeral 7.
Concept: Writing of numeral 7.

Exercise 7

Trace and write 8.

8

8

8

Using this page: Have students count the racing cars and trace the dotted line around the racing track. Then have them trace and practice writing numeral 8.
Concept: Writing of numeral 8.

Trace and write 9.

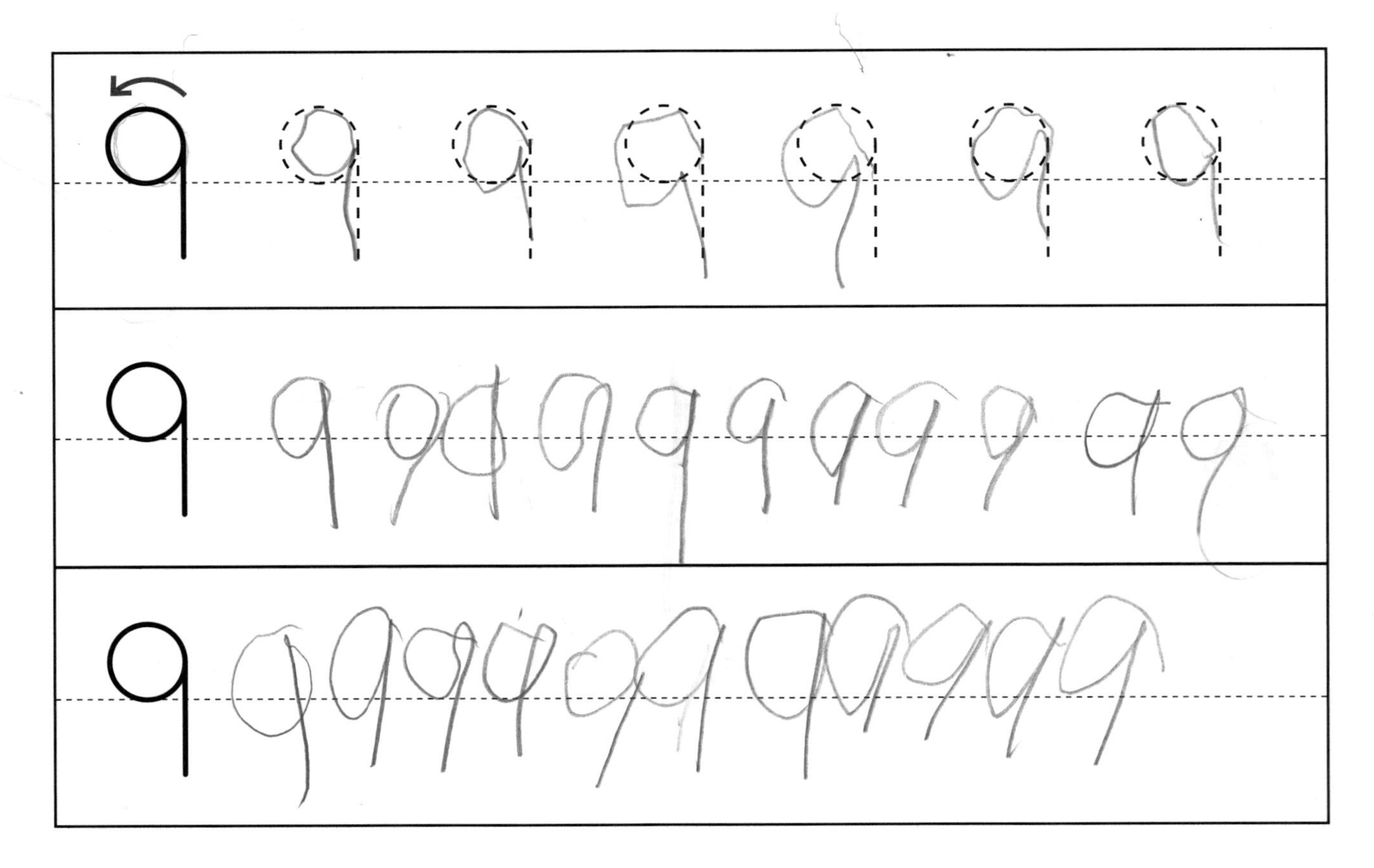

Using this page: Have students count the dinosaurs and trace the dotted lines around the biggest dinosaur's head. Then have them trace and practice writing numeral 9.
Concept: Writing of numeral 9.

Trace and write 10.

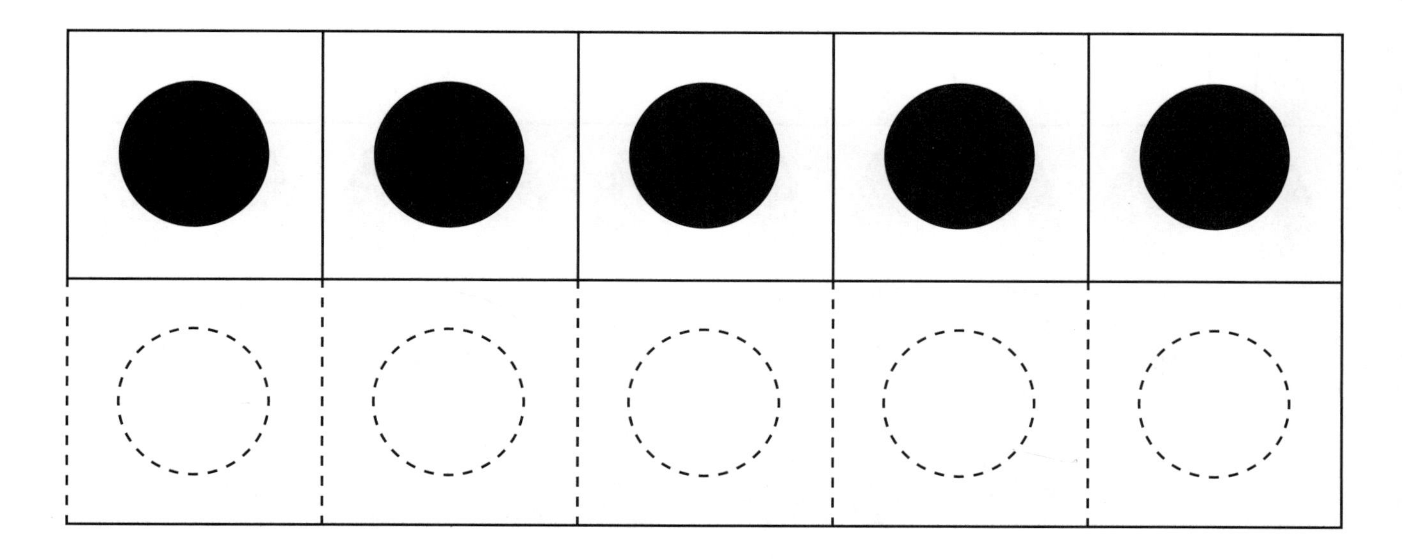

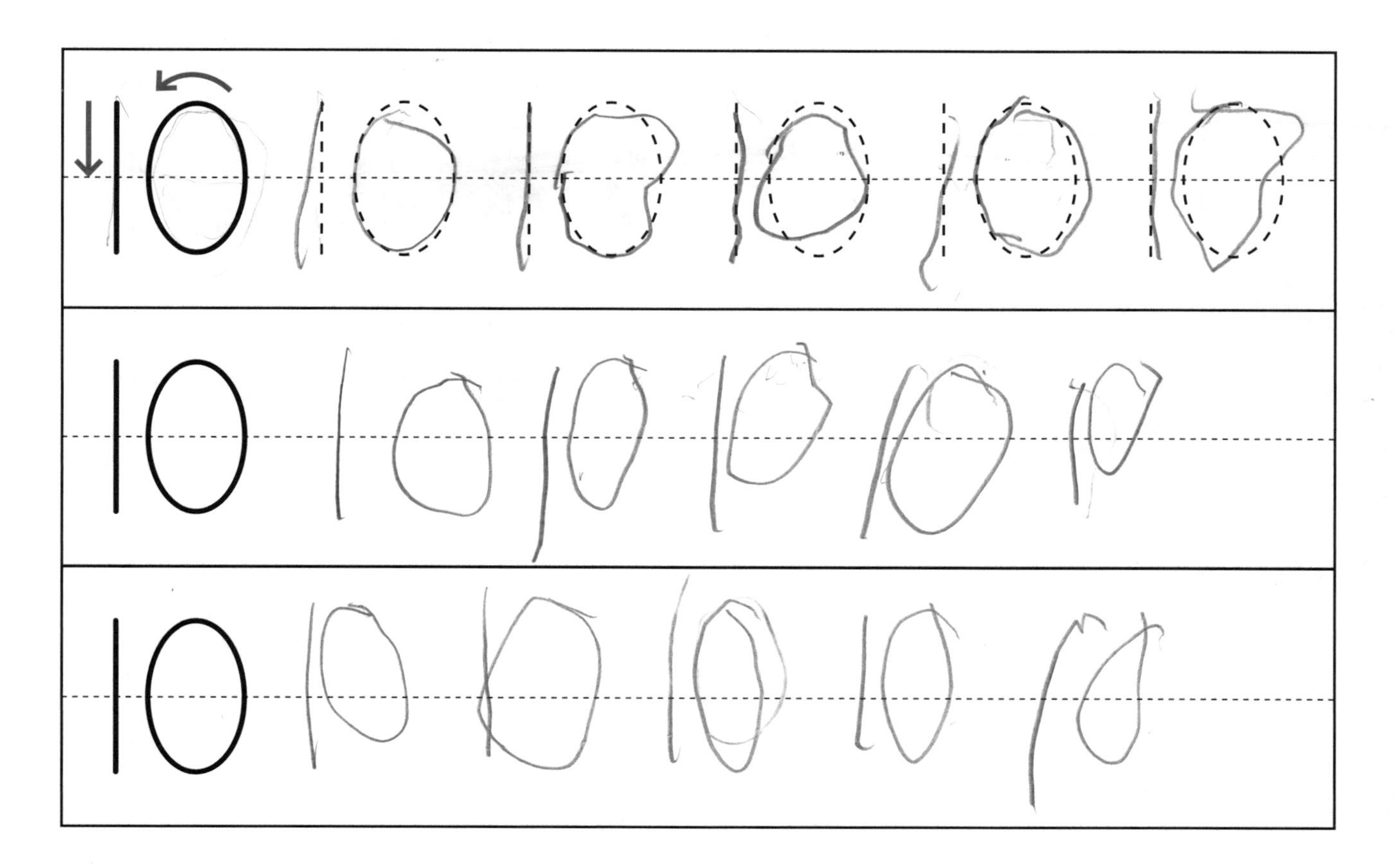

Using this page: Have students trace the dotted lines and and count the number of dots in the ten-frame. Then have them trace and practice writing numeral 10.
Concept: Writing of numeral 10.

Write the missing numbers in the boxes.

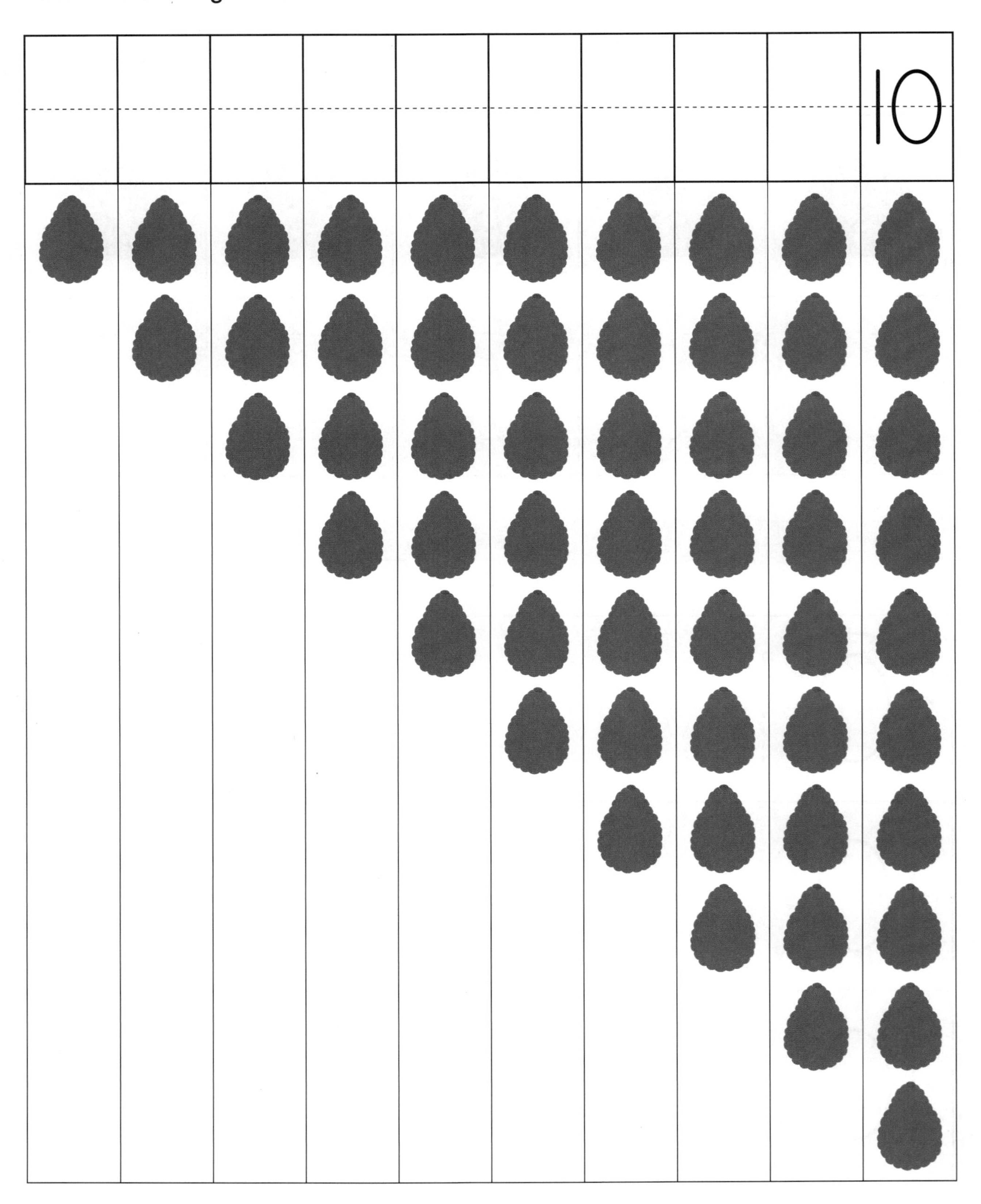

Using this page: Have students look at the number path and write the numerals in the boxes. Have them count the avocados if they do not know what the missing number is.
Concept: Numeral sequence and writing of 1 to 10.

Exercise 8

Write the number in the box.

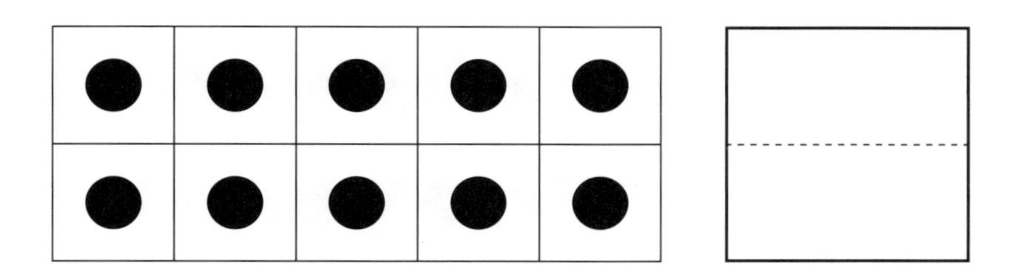

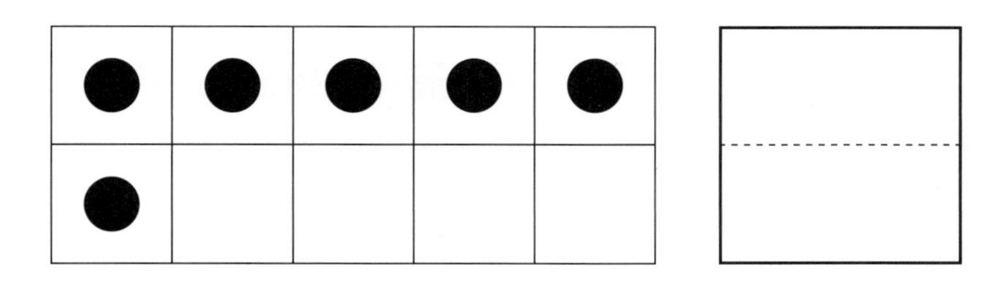

Using this page: Have students look at the number of dots on each ten-frame card and write the numeral in the box.
Concept: Writing of numeral 6 to 10.

Count and write the number.

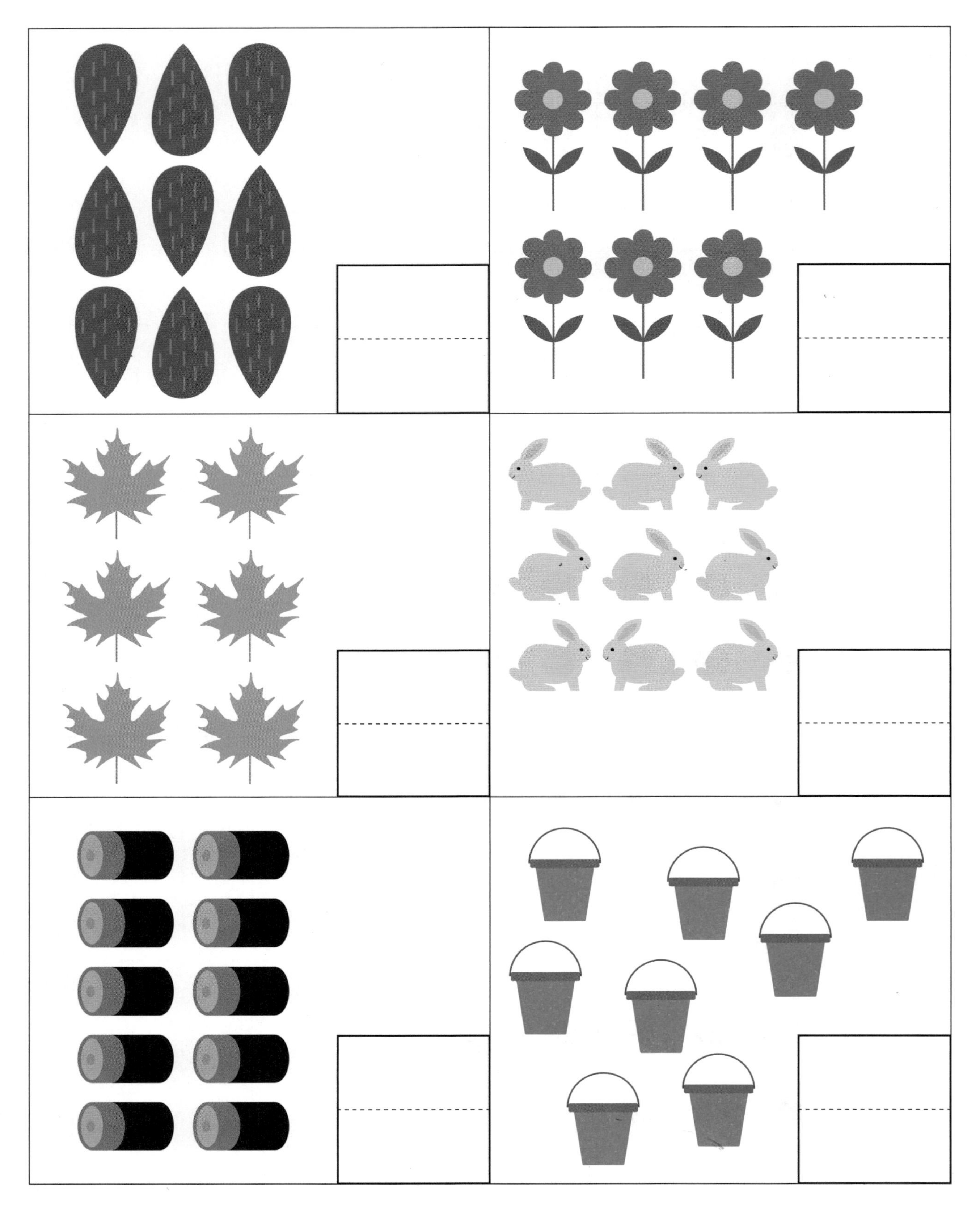

Using this page: Have students count and write the number of objects in the box.
Concept: Counting and writing of numerals 6 to 10.

Exercise 9

Write the number in the box.

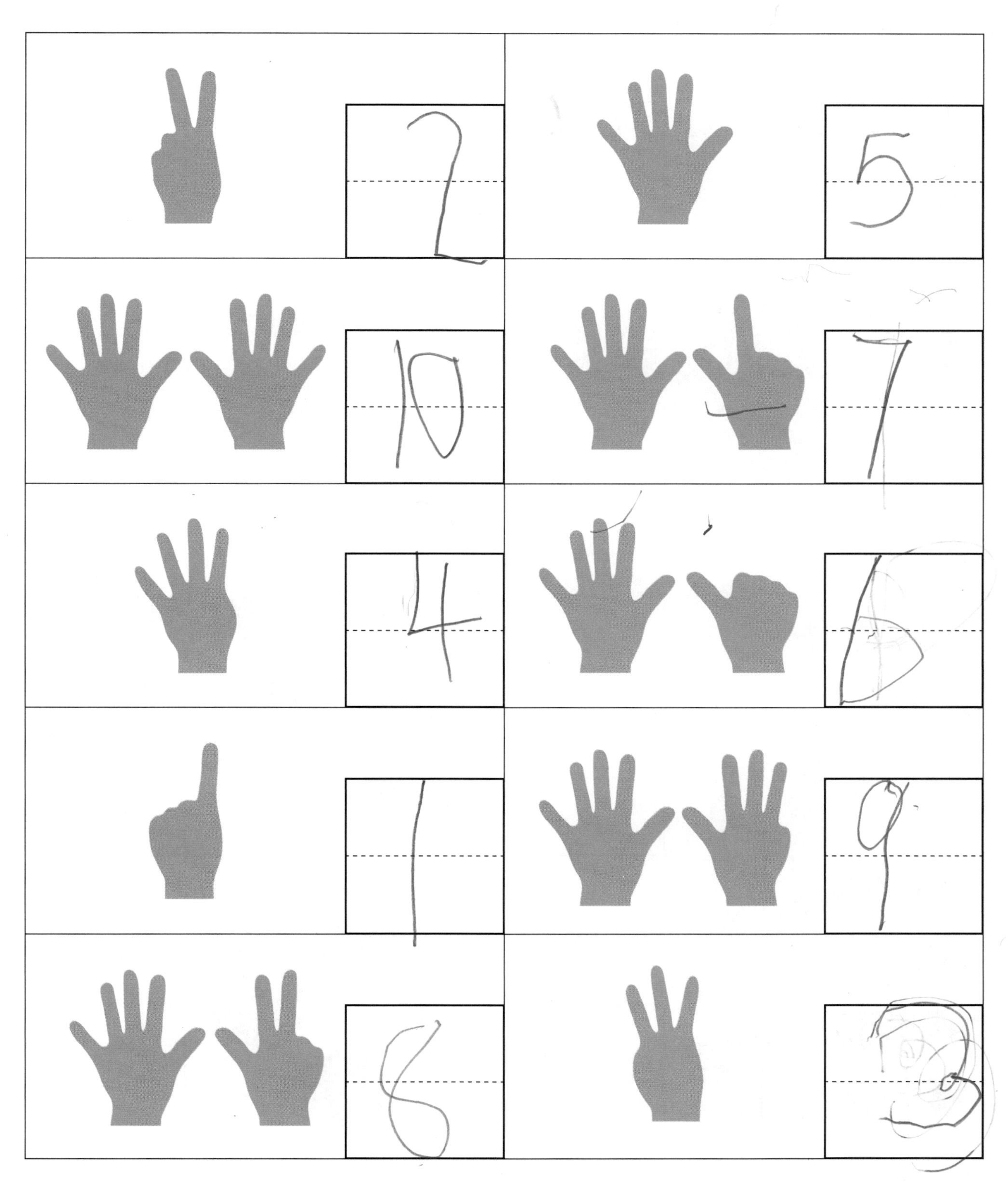

Using this page: Have students count and write the number of fingers in the box.
Concept: Writing of numerals 1 to 10.

Count and write the number.

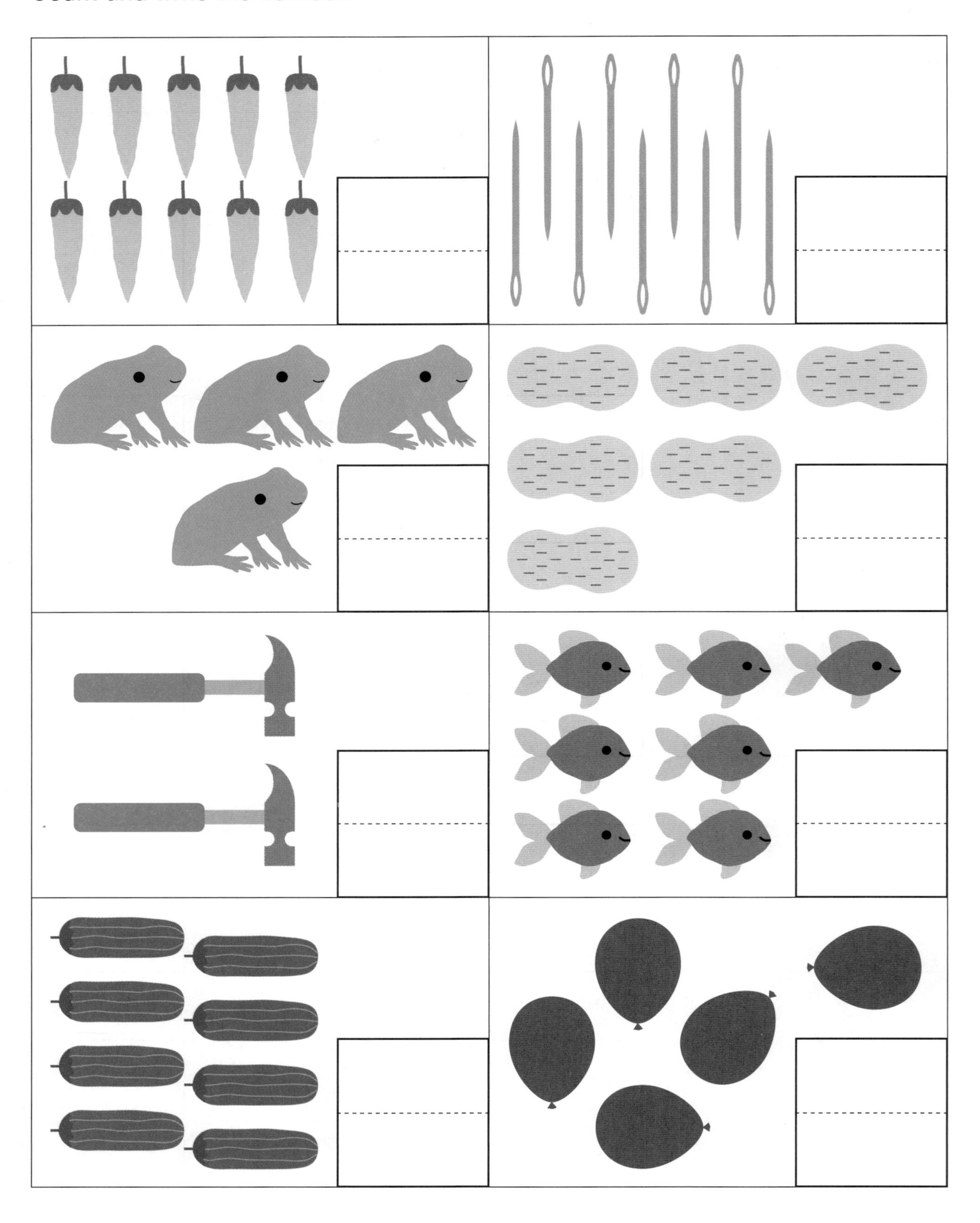

Using this page: Have students count and write the number of objects in the box.
Concept: Counting and writing of numerals 1 to 10.

Write the number in the box.

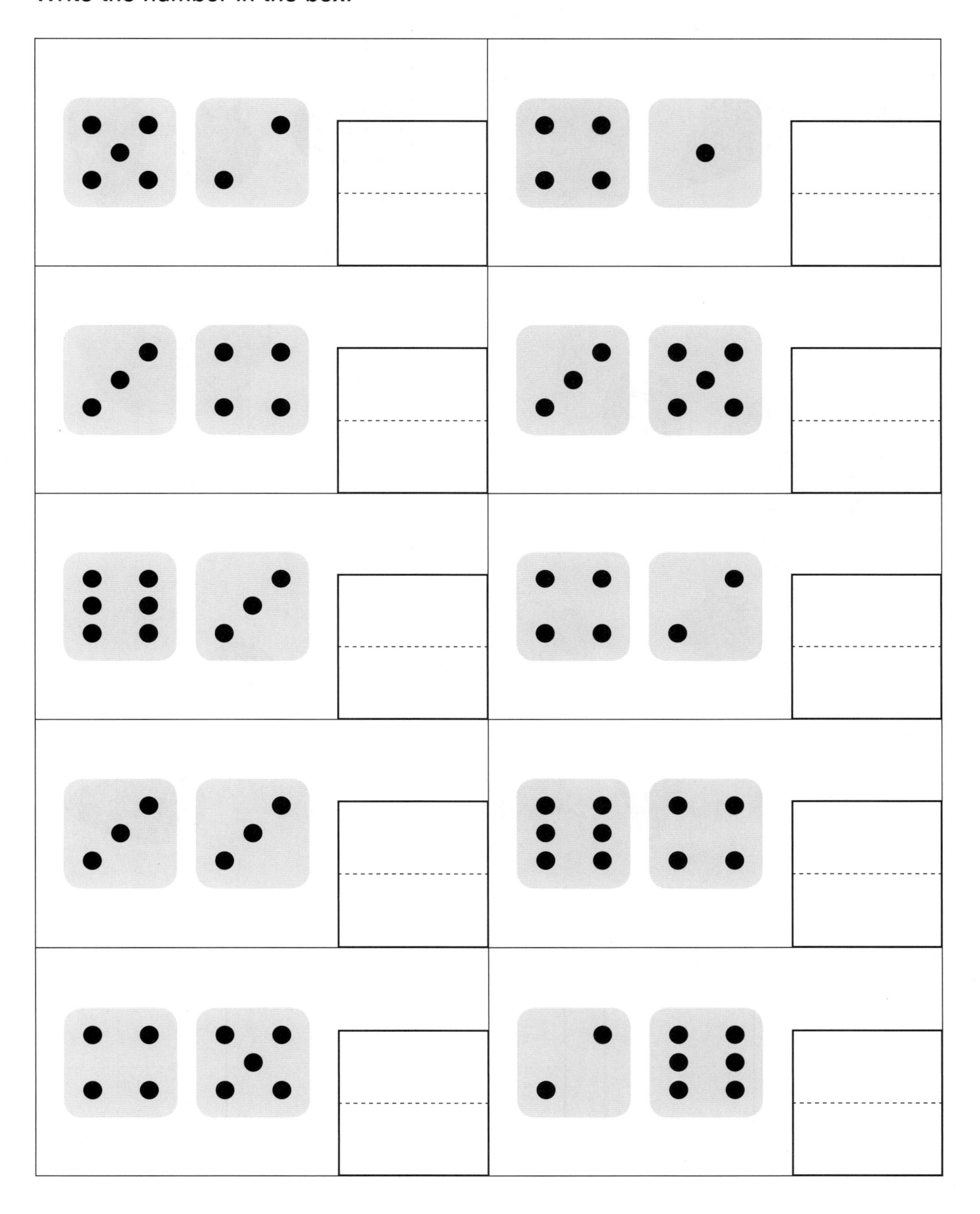

Using this page: Have students count and write the number of dots in the box.
Concept: Counting and writing of numerals 1 to 10.

Find, count, and write the number.

Using this page: Have students count the number of objects for each color and write the numeral in the box.
Concept: Writing of numerals 1 to 10.

Exercise 10

The first has a ▲.

Count from the front and circle the correct animal.

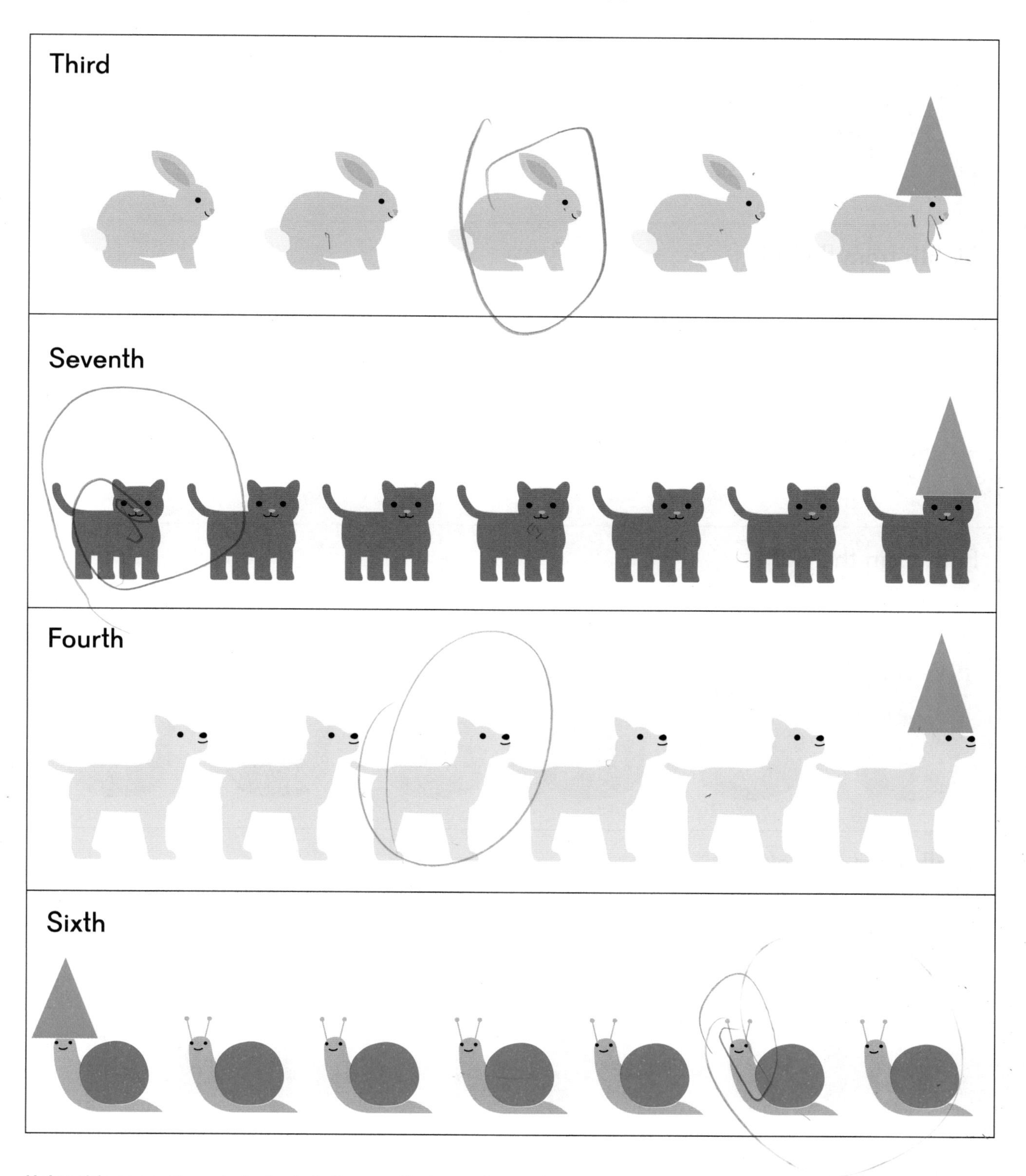

Using this page: Have students circle the specified animal in each row.
Concept: Ordinal positions first through tenth from the front.

Follow the directions and circle.

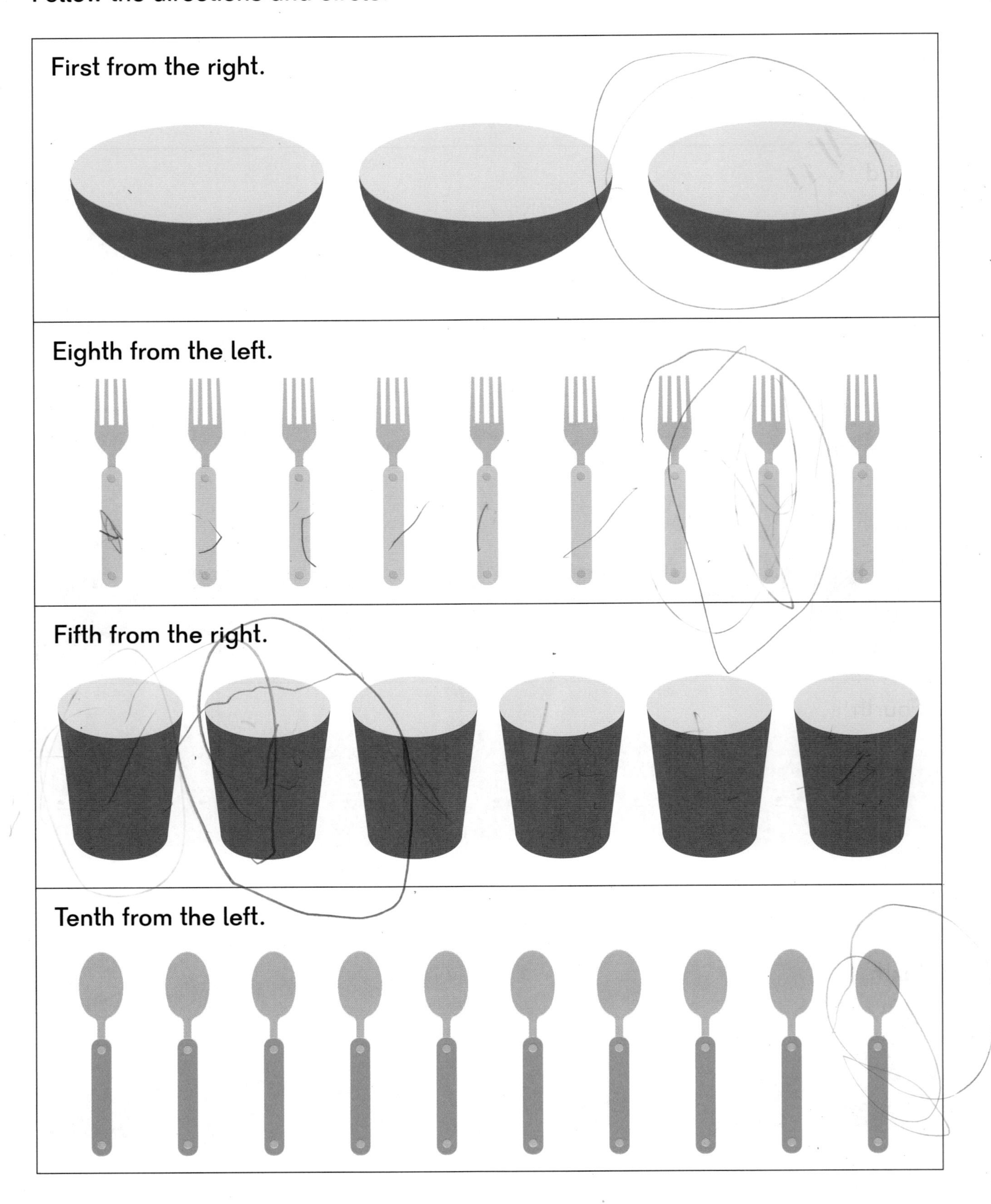

Using this page: Have students circle the specified object in each row.
Concept: Ordinal positions first through tenth from the left or right.

Count from the top and match.

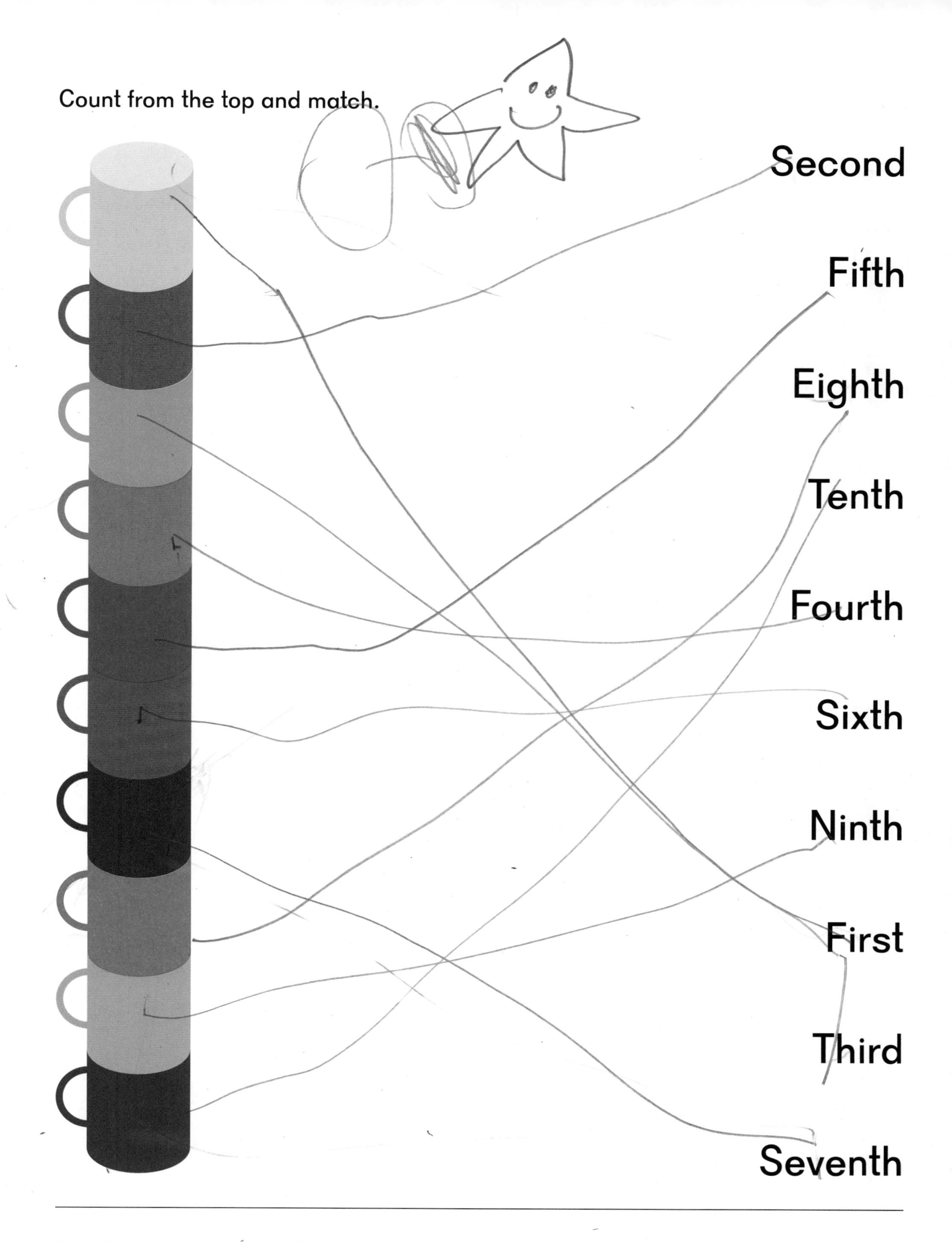

Using this page: Have students follow the directions and match the ordinal numbers to the specified cup.
Concept: Ordinal positions first through tenth from the top.

Count from the bottom and color according to the Color Key.

Color Key
First
Fourth
Sixth
Ninth
Seventh
Third

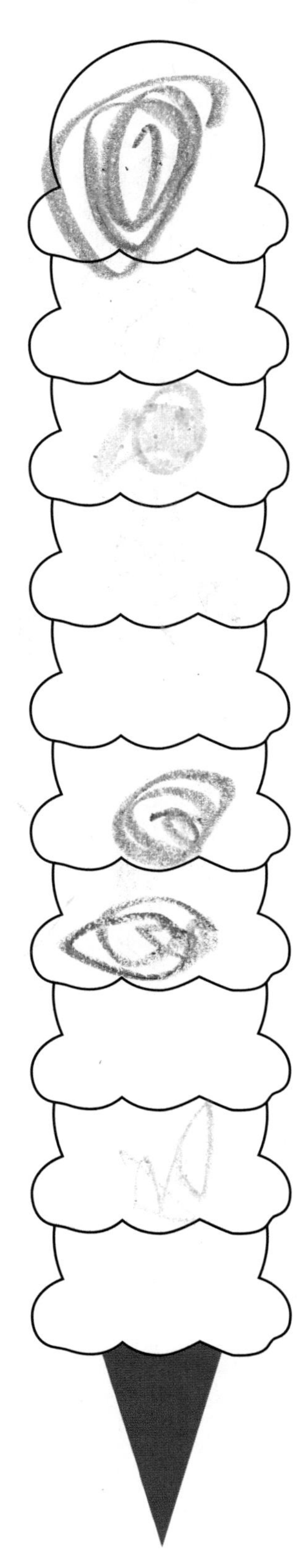

Using this page: Have students follow the directions and color the ice cream.
Concept: Ordinal positions first through tenth from the bottom.

Exercise 11

Circle the group that has 1 more in each box.

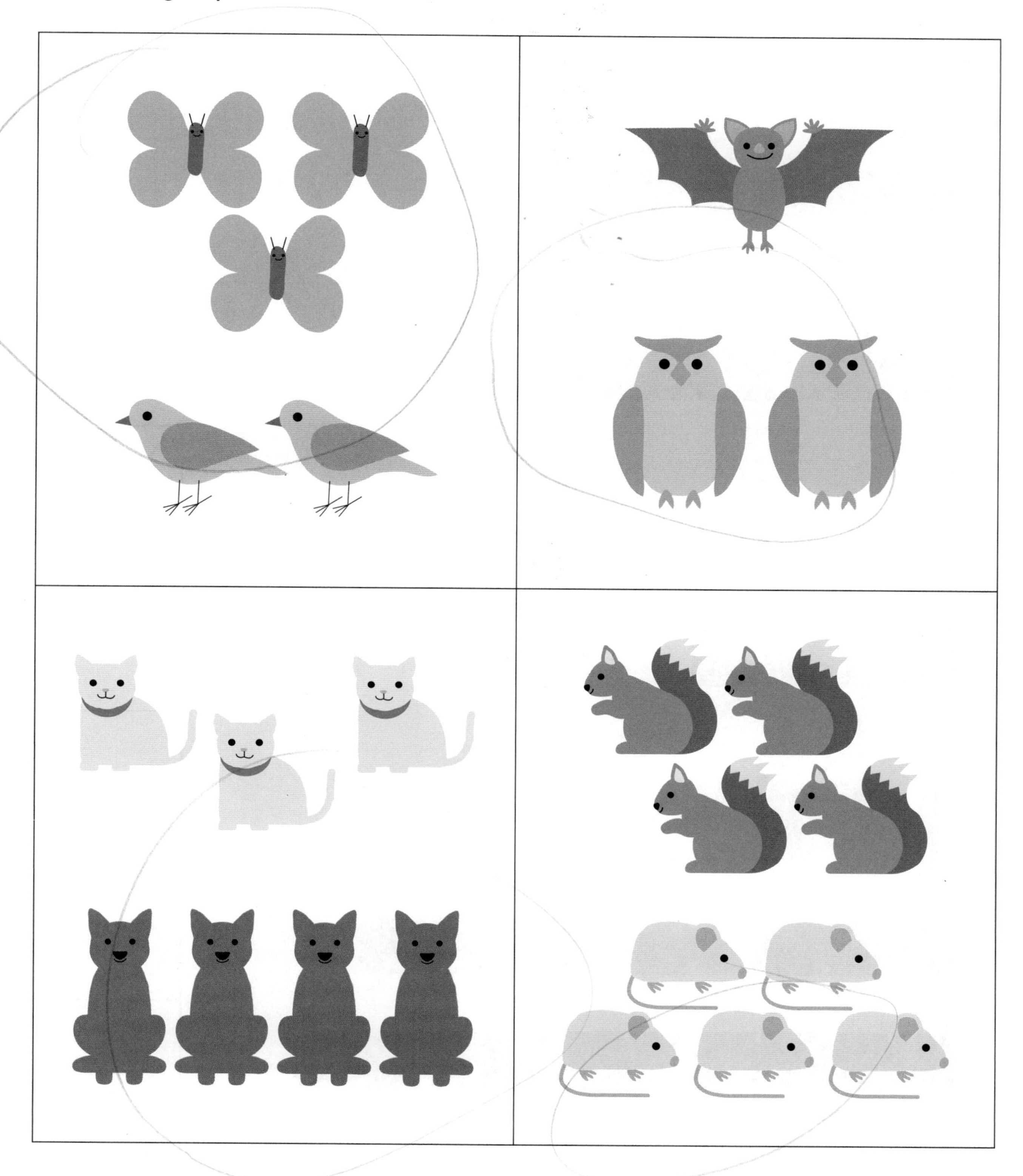

Using this page: Have students compare the two groups of objects in each box and circle the group that has one more than the other group.
Concept: Each successive number name refers to a quantity that is one more than the previous quantity.

Count and write the numbers.
Circle the number that is 1 more.

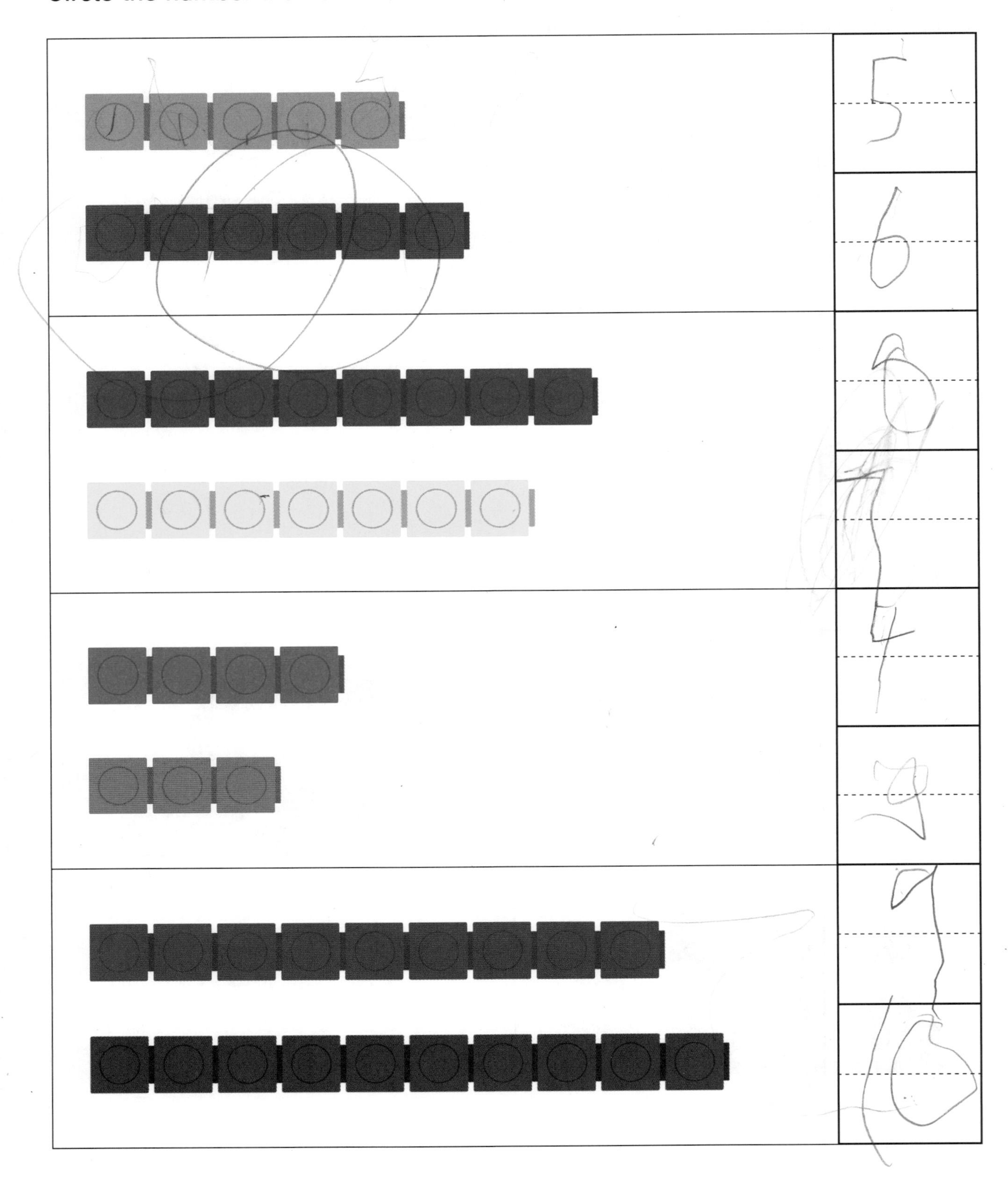

Using this page: Have students count each group of objects and write the numeral in the box. Then have them circle the number that is one more than the other.
Concept: Each successive number name refers to a quantity that is one more than the previous quantity.

Circle the group that has 1 more than the number.

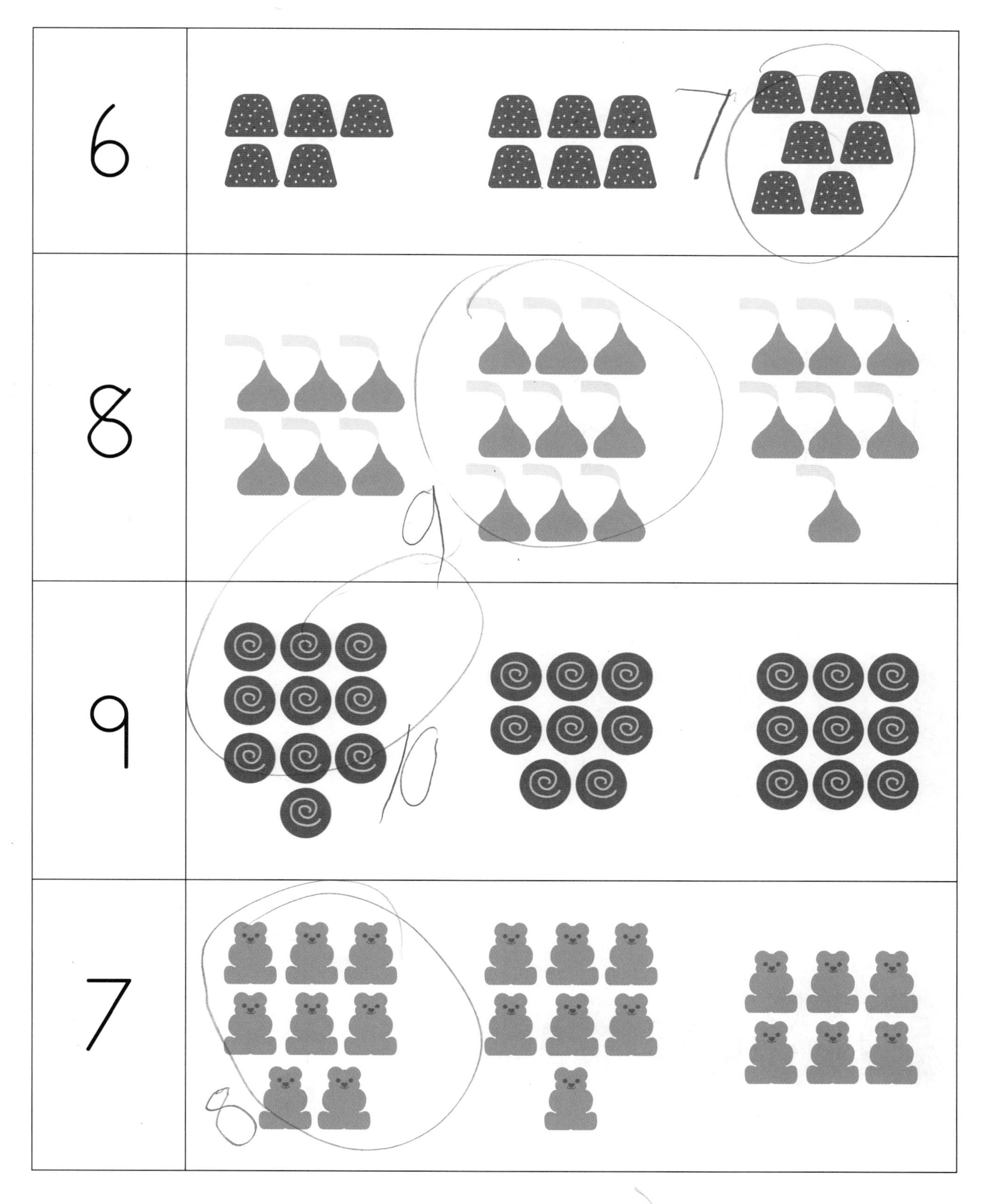

Using this page: Have students look at the numeral in each row and find the group of objects that has one more than the number, then circle it.
Concept: Each successive number name refers to a quantity that is one more than the previous quantity.

Circle the number that comes next.

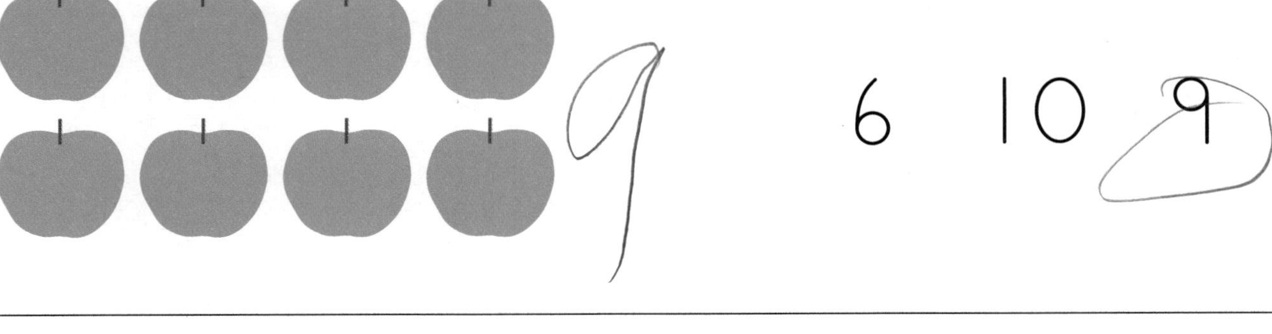

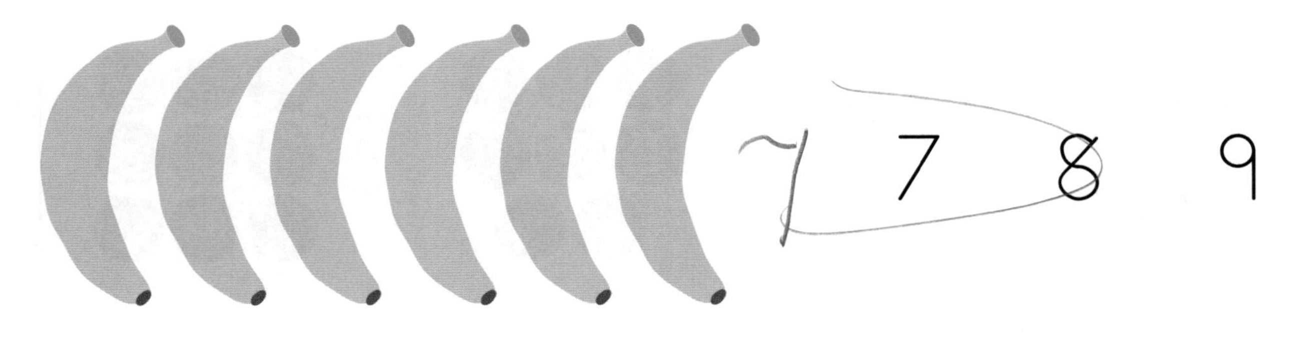

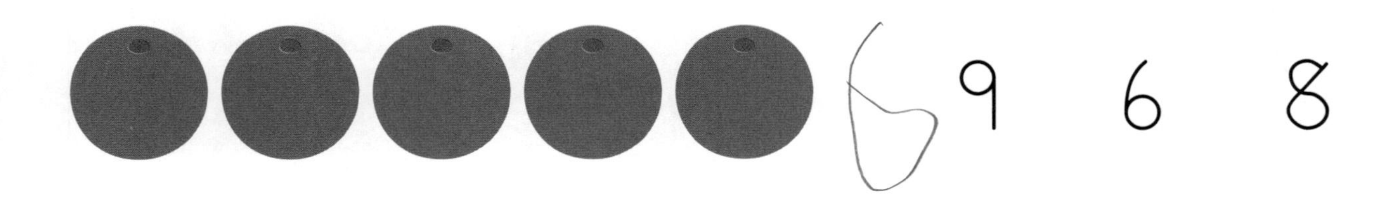

Using this page: Have students count the objects to find the next number when counting on, then circle the correct numeral.
Concept: Each successive number name refers to a quantity that is one more than the previous quantity.

Exercise 12

Connect the dots.

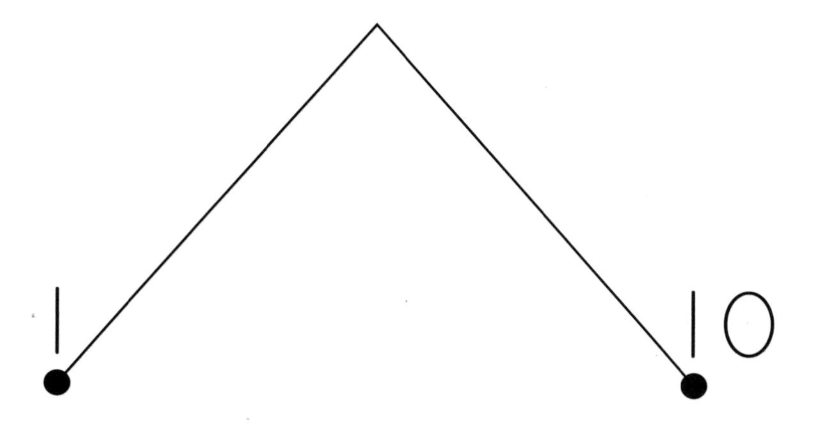

2 3 8 9

4

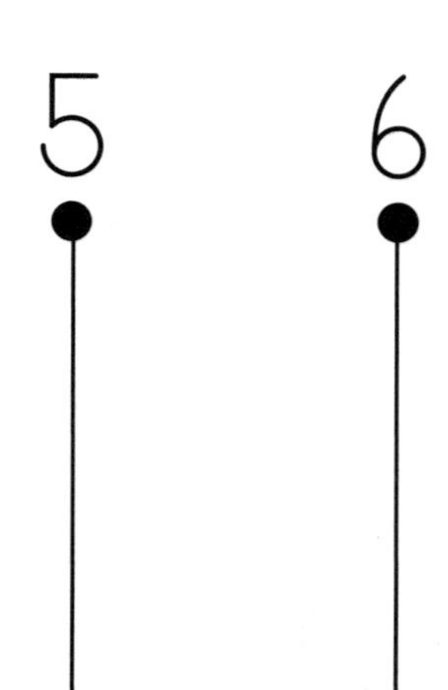

7

Using this page: Have students connect the dots in the number sequence from 1 to 10 to complete the picture.

Write the number in the box.

Using this page: Have students count the number of fingers, then write the numeral in the box.

Match the number with the group that has 1 more.

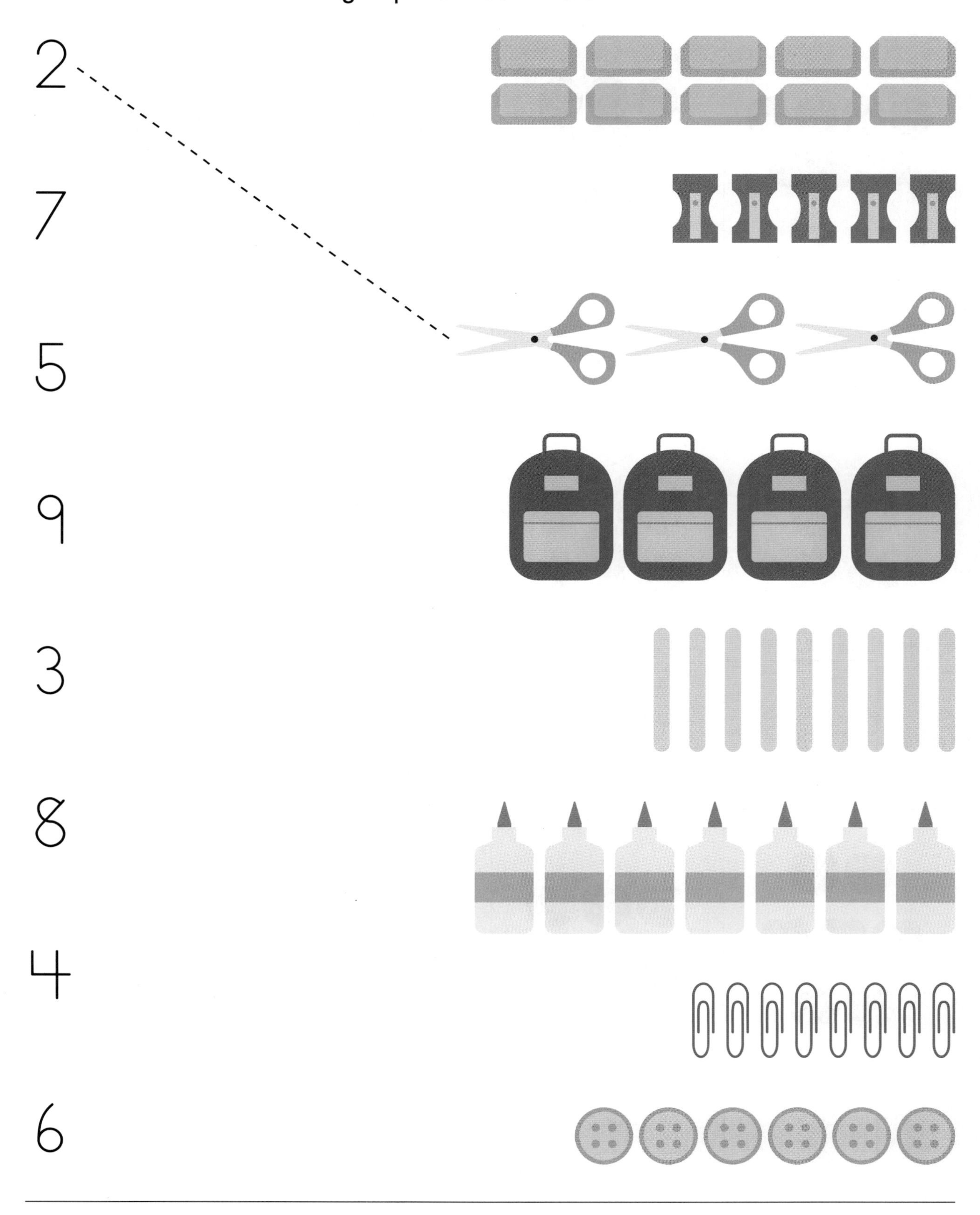

Using this page: Have students look for the group of objects that has one more than the numeral and match.

Count, write, and circle the group that comes next.

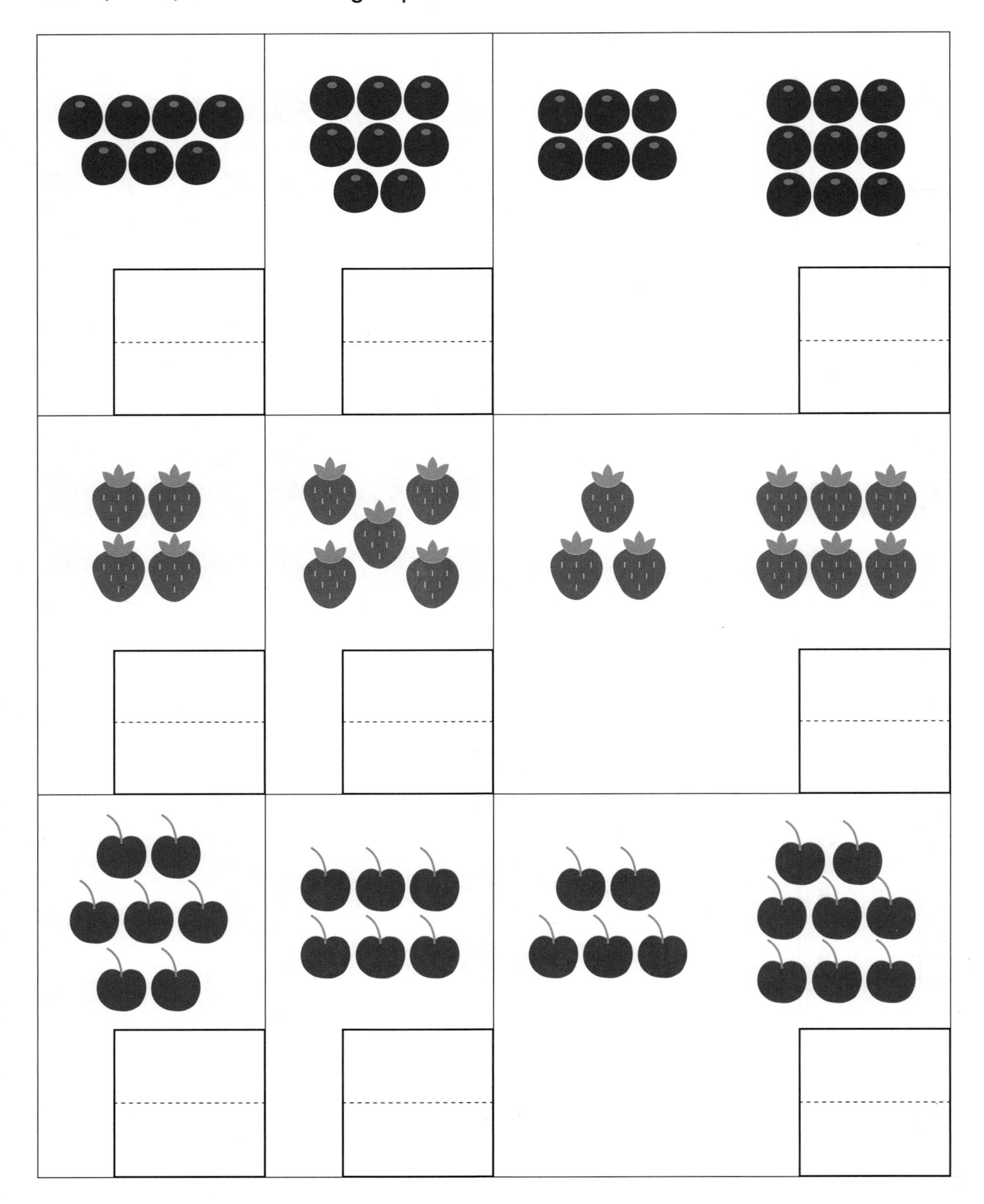

Using this page: Have students count the number of objects in the first two groups and write the numeral in the box. Then, have them write the numeral that comes next in the sequence and circle the correct group of objects.

Ordinal

Follow the directions and color the fence.

Color the fifth from the left ●.

Color the seventh from the right ●.

Color the ninth from the left ●.

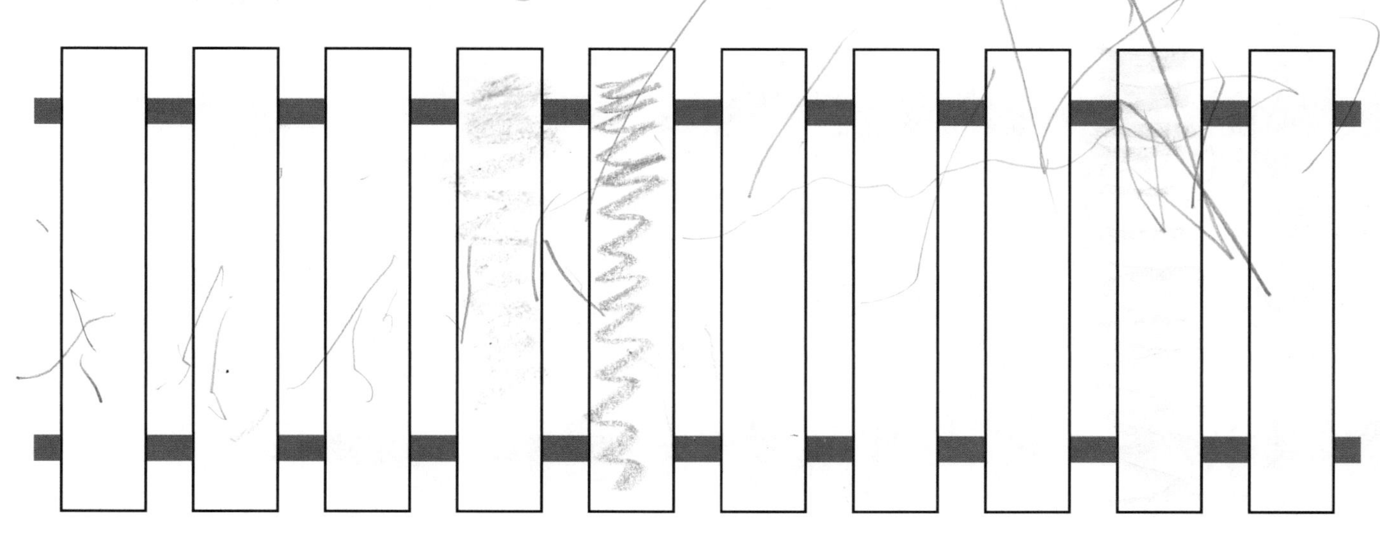

Follow the directions and color the 6 cubes.

Color the second from the top ●.

Color the third from the bottom ●.

Color the sixth from the top ●.

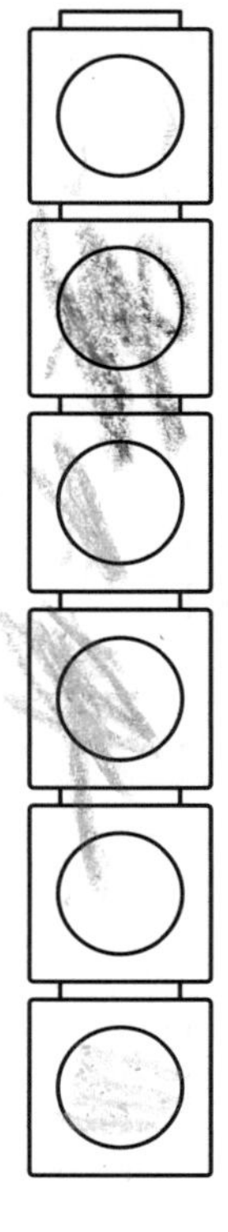

Using this page: Have students color the specified object in each row.

Concept: Ordinal positions first through tenth.

Find, count, and write the number.

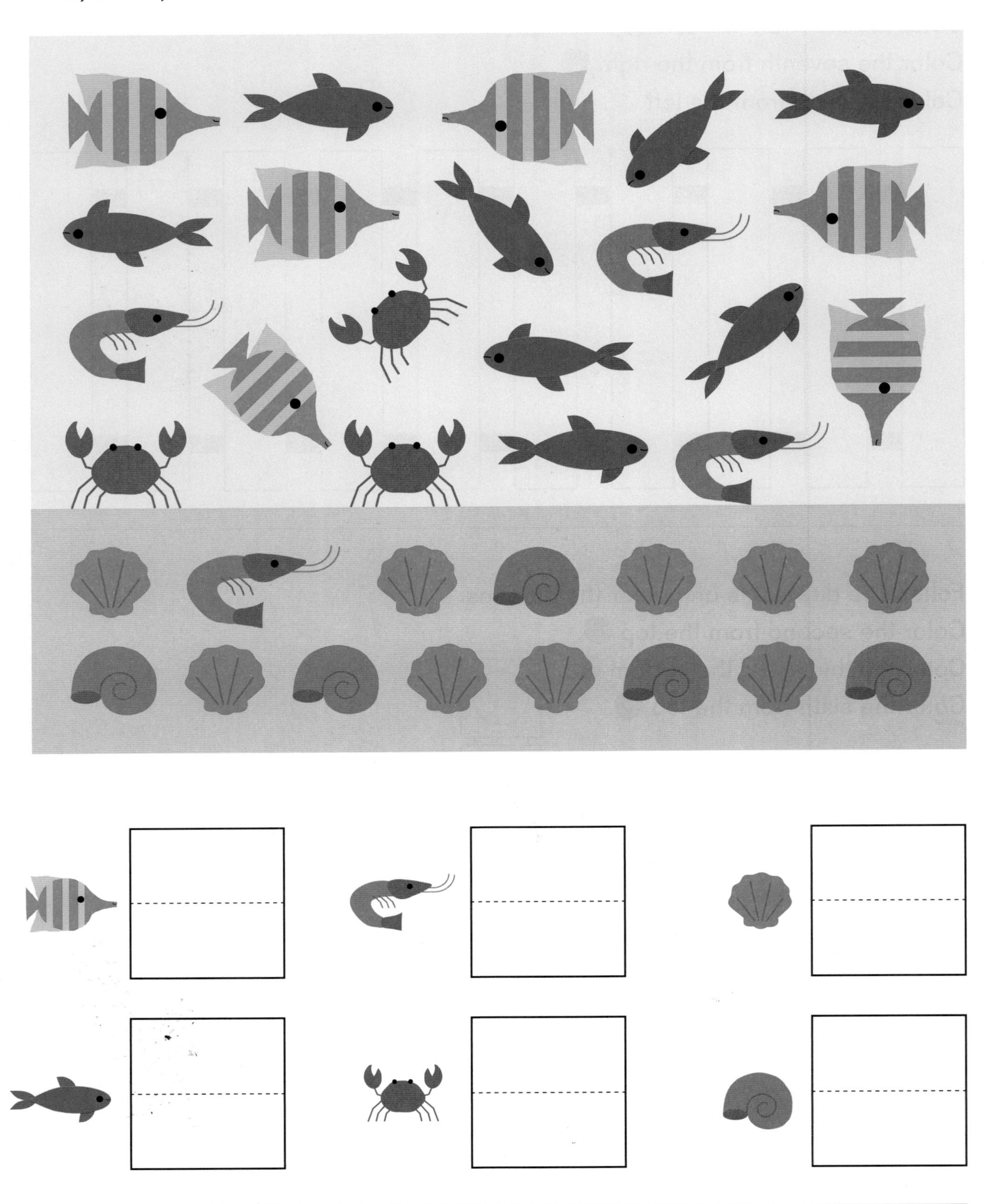

Using this page: Have students count the number of each object and write the numeral in the box.

Chapter 4 Shapes and Solids

Exercise 1

Color the solids that are like the first one.

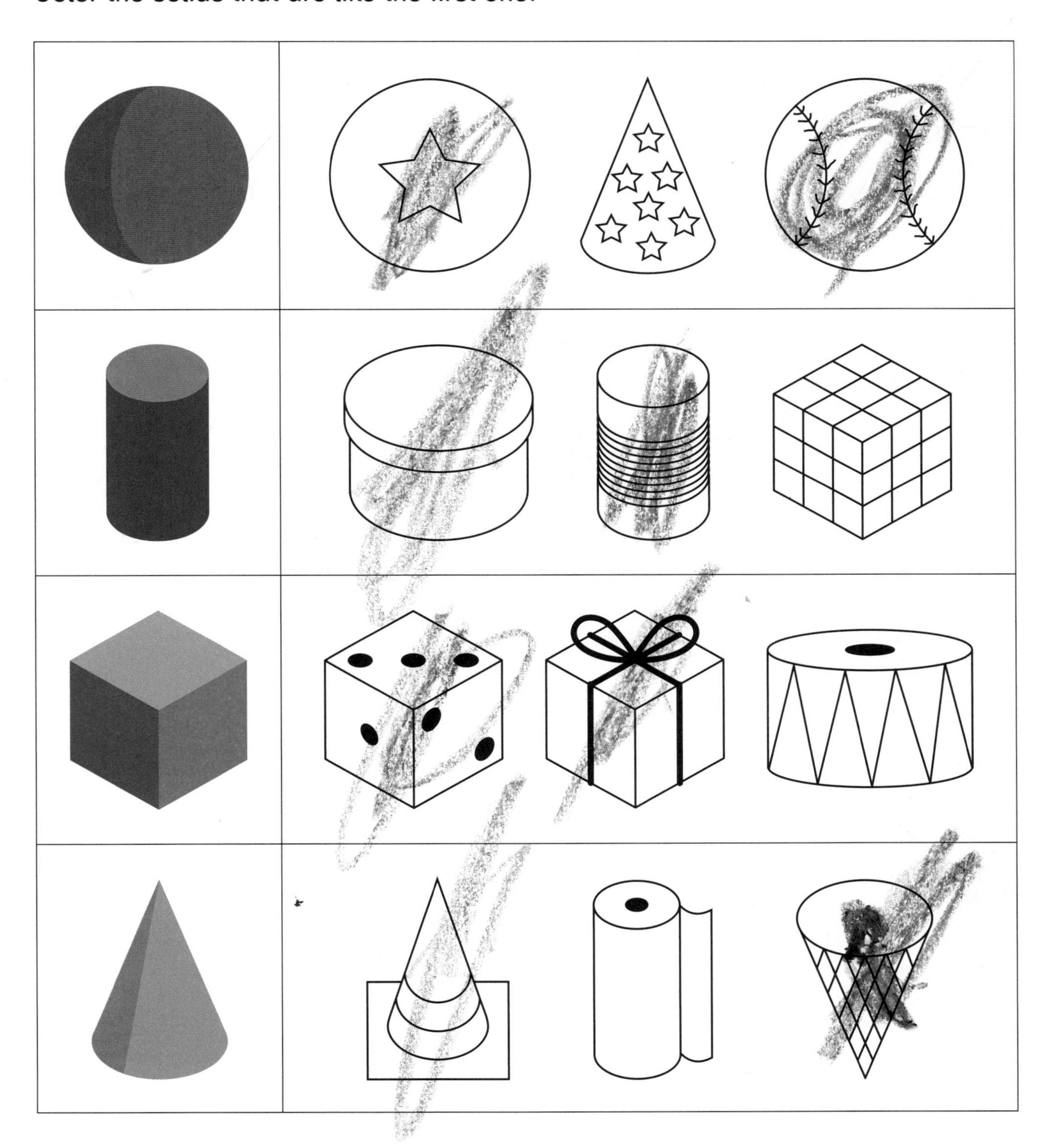

Using this page: Have students color the solids that are the same in each row.
Concept: Recognizing cubes, cylinders, spheres, and cones.

Color according to the Color Key.

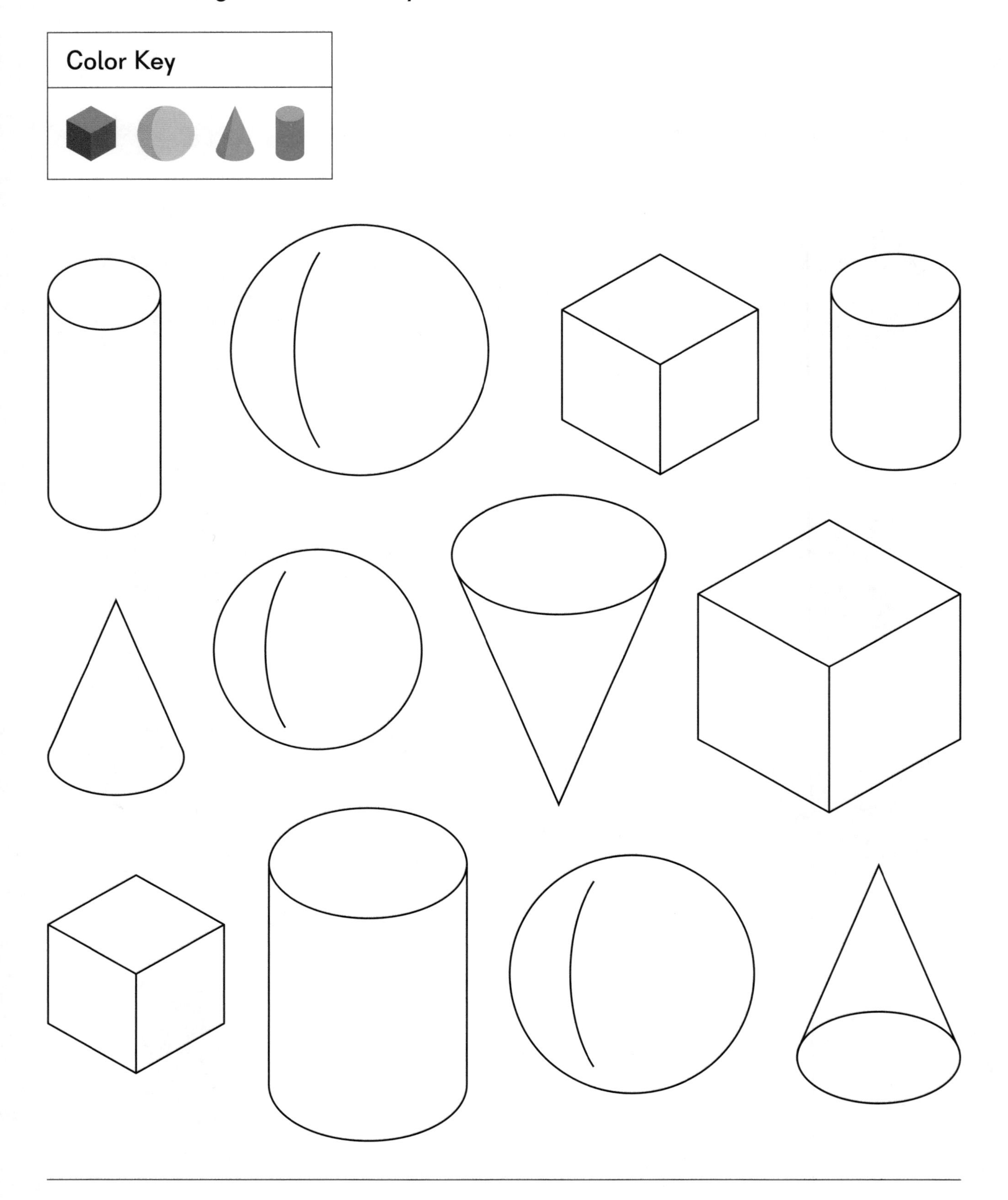

Using this page: Have students refer to the color key to color the solids.
Concept: Recognizing cubes, cylinders, spheres, and cones.

Exercise 2

Color the closed shapes.

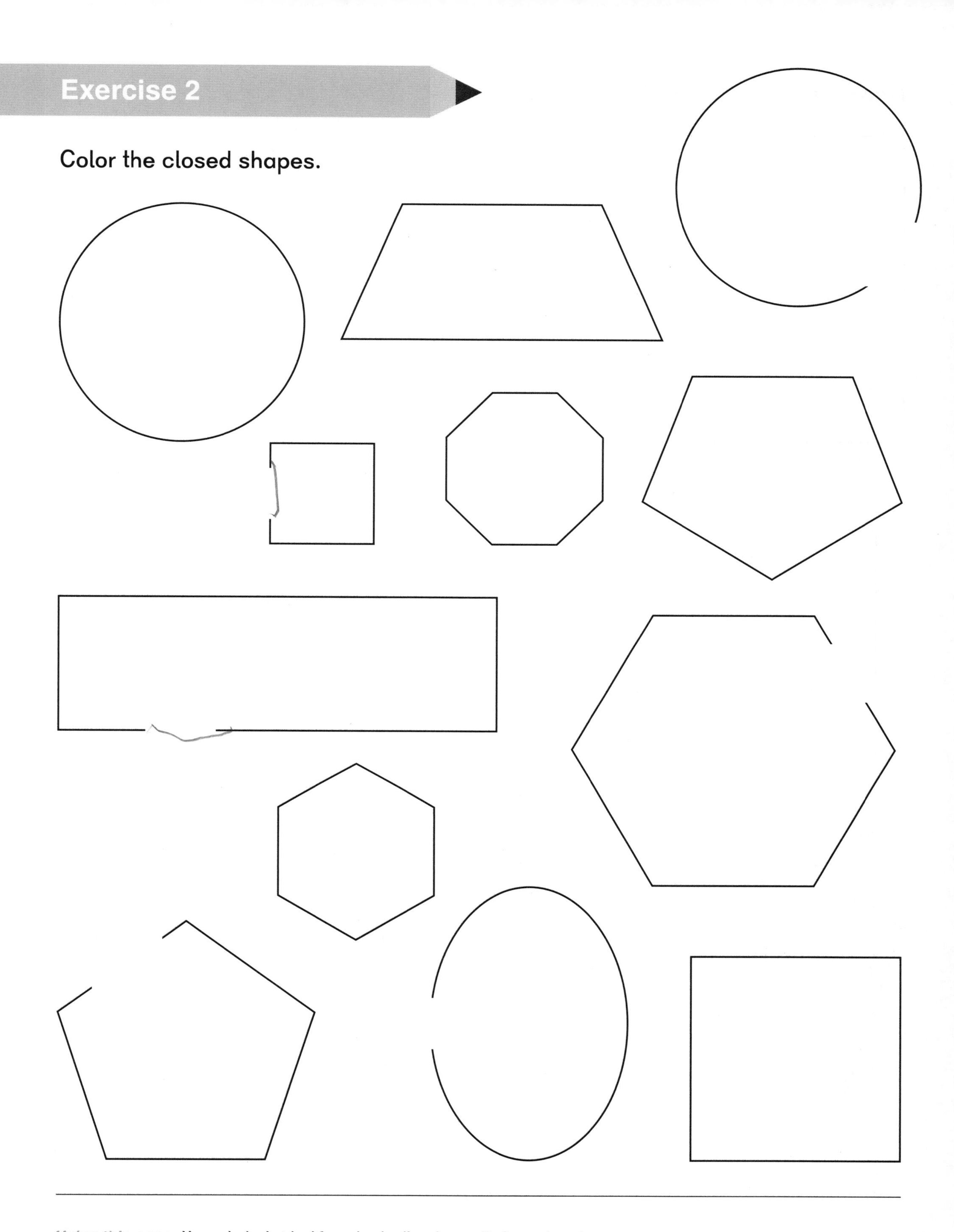

Using this page: Have students identify and color the shapes that are closed.
Concept: Identifying closed shapes.

Draw lines to close the shapes.

Using this page: Have students add one or two lines to each open shape to make a closed shape.
Concept: Distinguishing open and closed shapes.

Exercise 3

Color the rectangles.

Using this page: Have students identify the rectangles and color them.
Concept: Identifying rectangles.

Match.

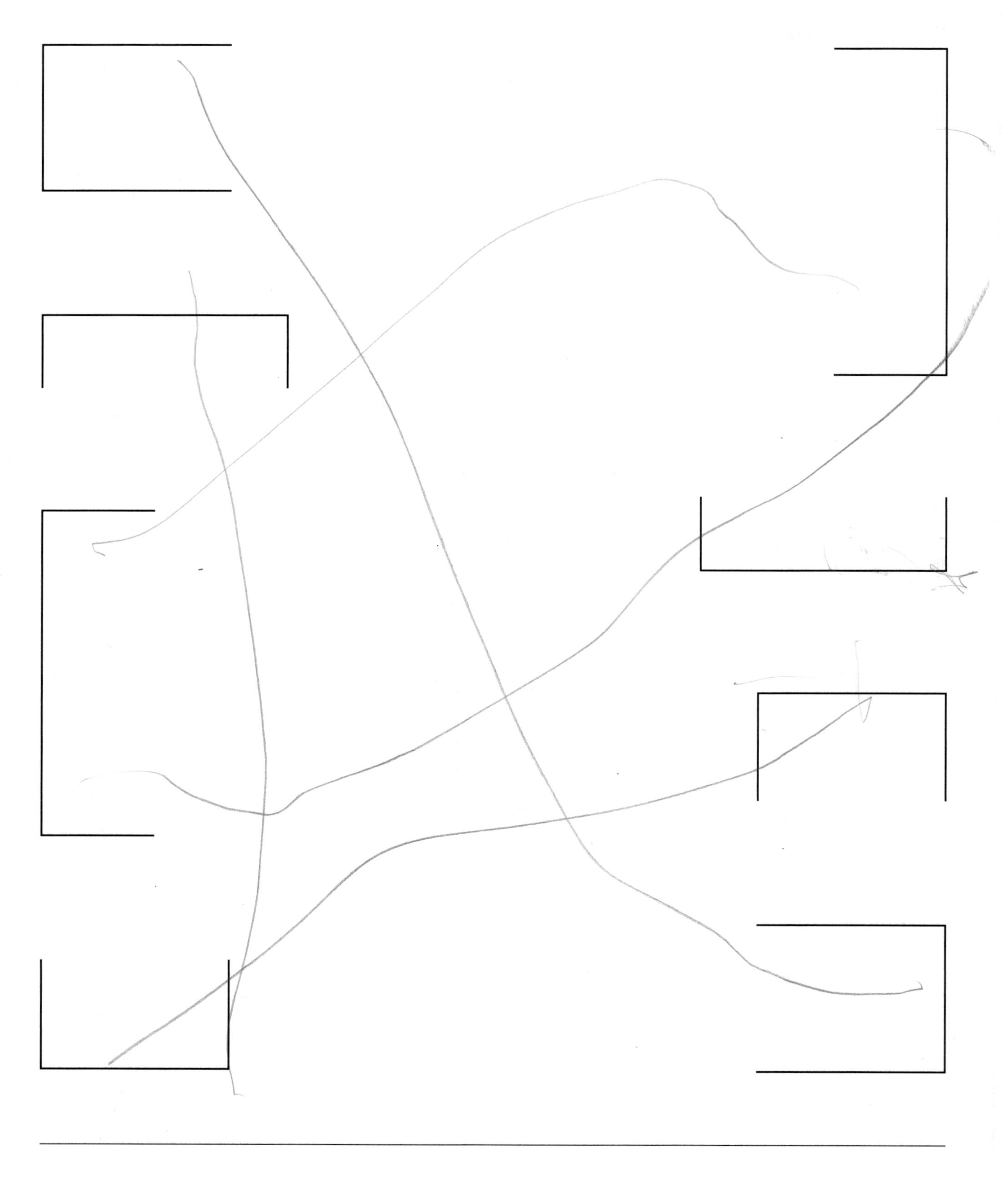

Using this page: Have students look for the other half of each rectangle and match.
Concept: Matching rectangles.

Exercise 4

Color the squares.

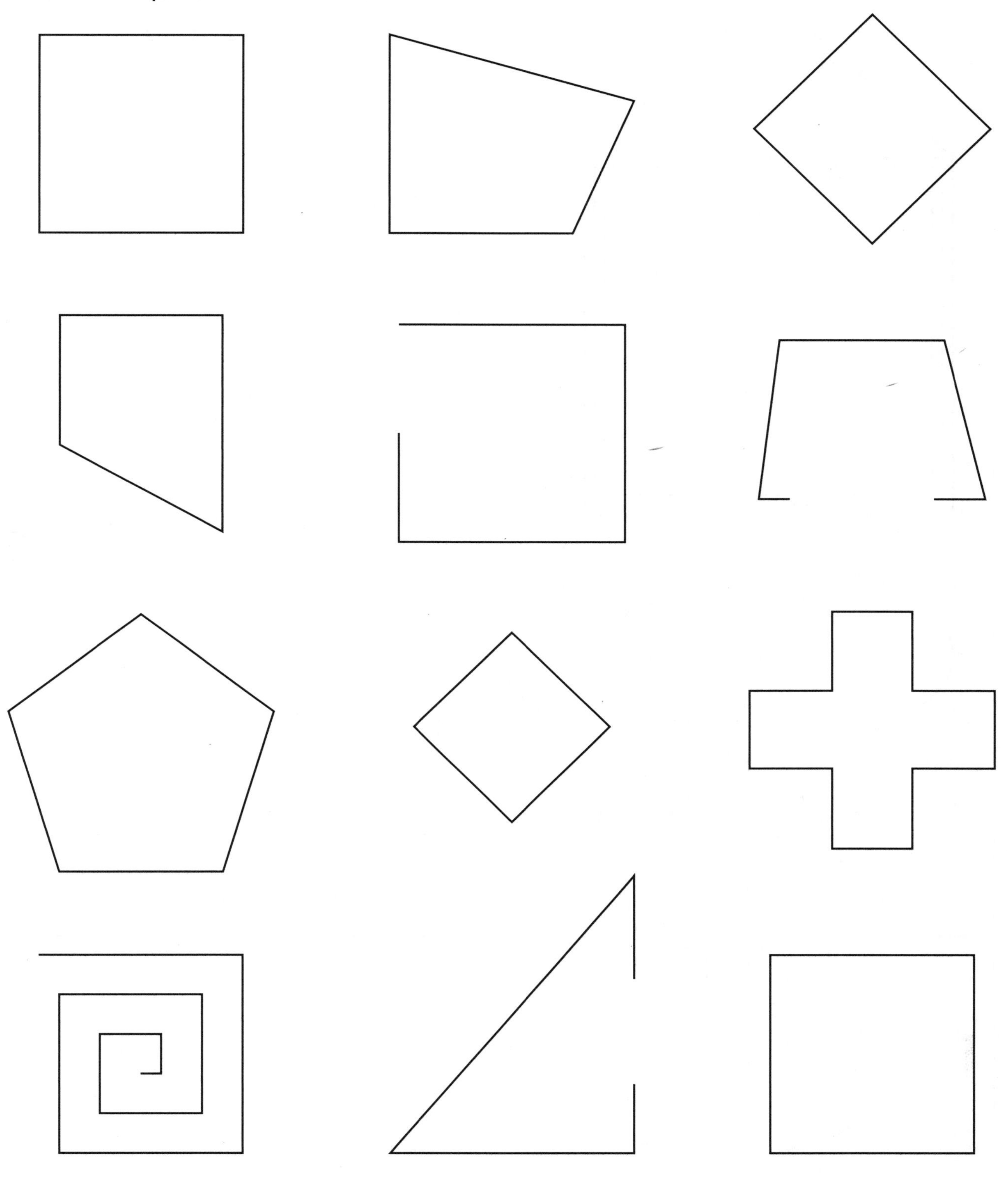

Using this page: Have students identify the squares and color them.
Concept: Identifying squares.

Color the squares.

Using this page: Have students identify the squares and color them.
Concept: Identifying squares.

Exercise 5

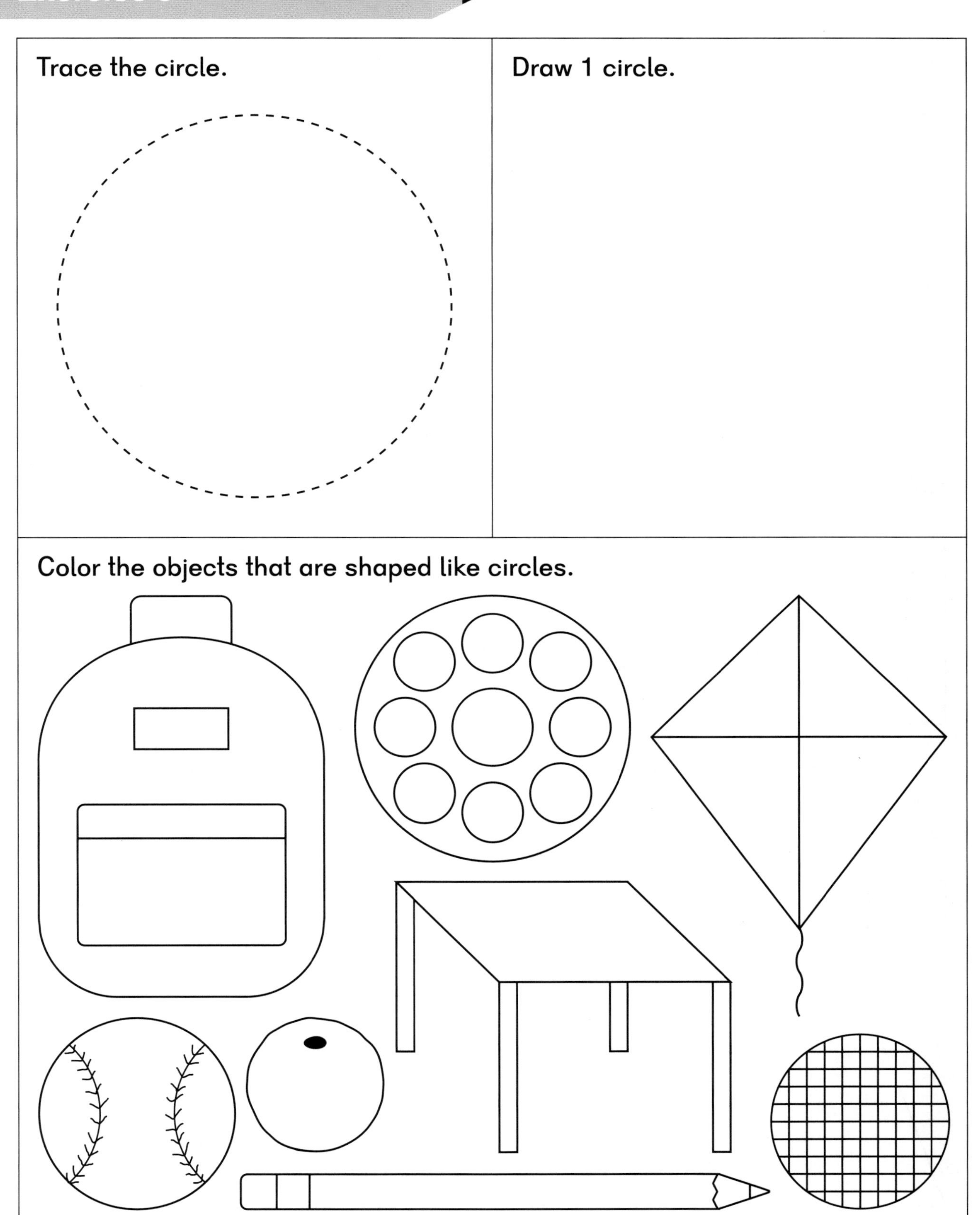

Using this page: Have students trace, draw, then identify the circles and color them.
Concept: Identifying circles.

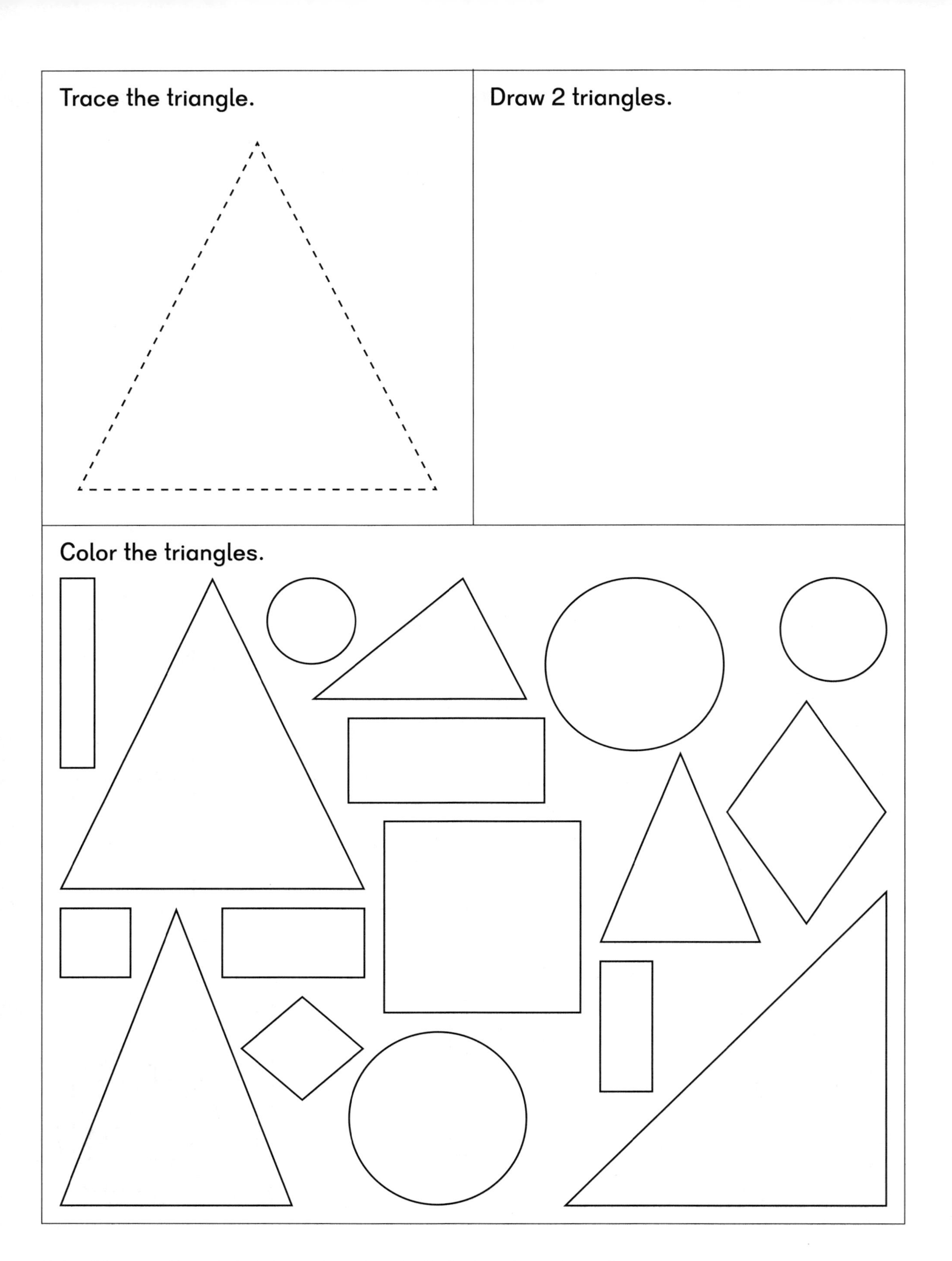

Using this page: Have students trace, draw, then identify the triangles and color them.
Concept: Identifying triangles.

Color according to the Color Key.

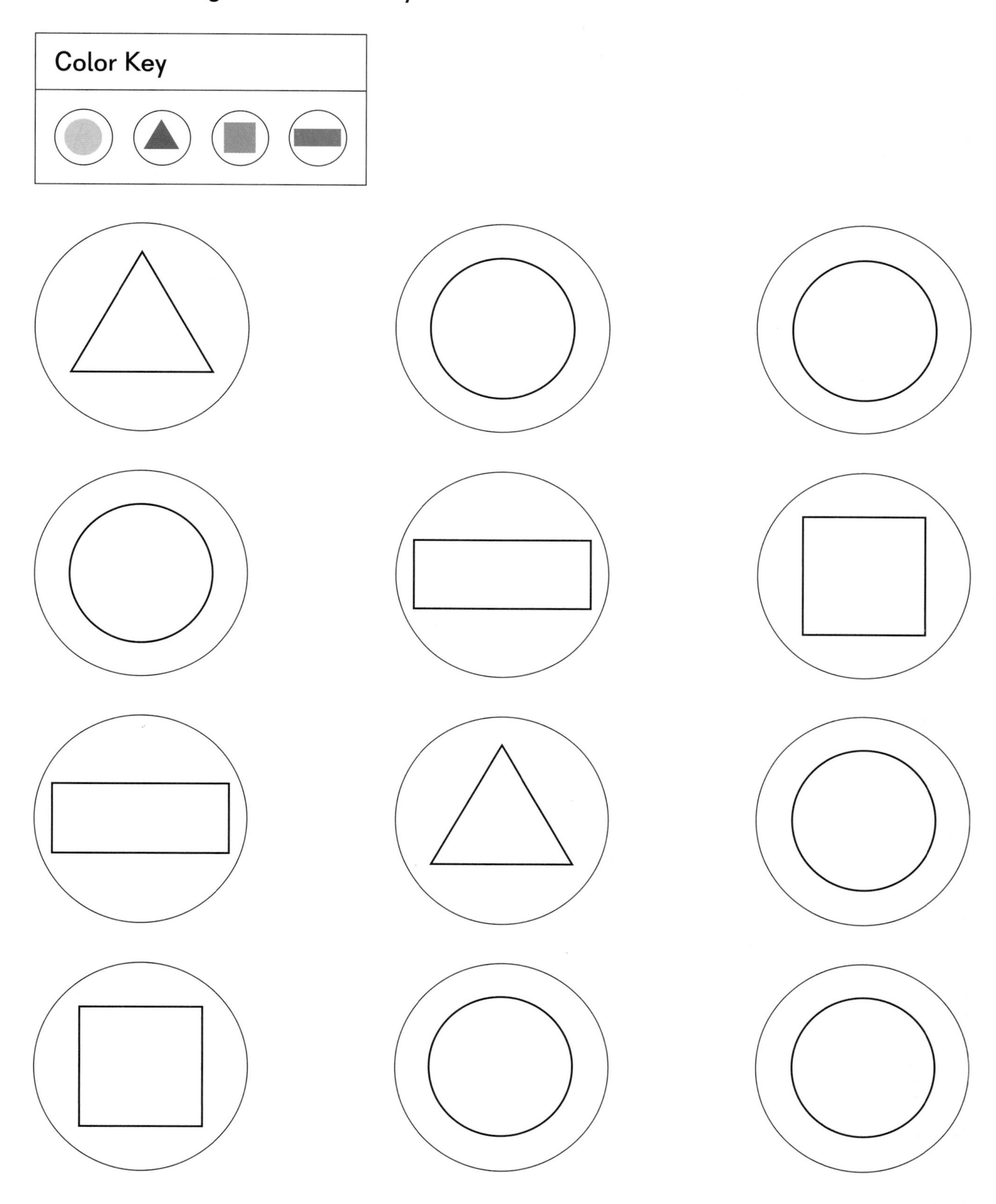

Using this page: Have students identify all the specified circles and color.
Concept: Identifying basic shapes.

Color the triangles.

Using this page: Have students identify all the triangles and color.
Concept: Identifying triangles.

Exercise 6

Follow the directions and color.

Color the cube at the top ●.

Color the cylinder below the red cube ●.

Color the sphere next to the green cylinder ●.

Color the cylinder between the cube and sphere ●.

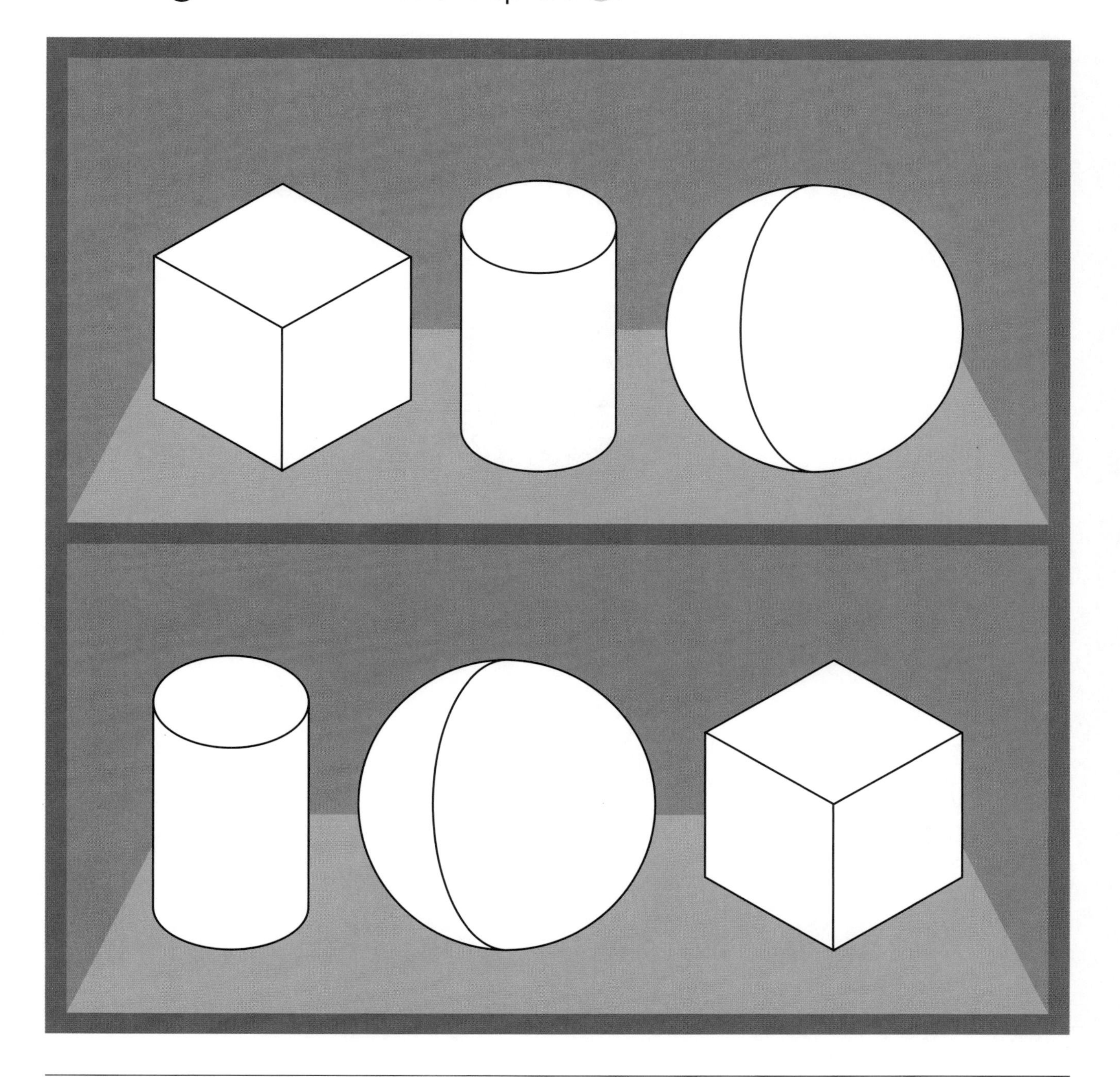

Using this page: Have students color the specified solids.
Concept: Identifying position of solids.

Follow the directions and paste the shapes.

Paste the ▲ above the ●.

Paste the ▬ below the ●.

Paste each ● on the ▬.

Paste the ▬ on the left of the ▬.

Paste the ▬ on the right of the ▬.

Paste the ■ below the ▬.

Before using this page: Pre-cut the shapes on page 147.
Using this page: Have students paste the cut-out shapes in the positions specified.
Concept: Identifying positions.

Exercise 7

Color the hexagons to form a path from the bee to the hive.

Using this page: Have students find and color all the hexagons that form a path from the bee to the hive.
Concept: Identifying hexagons.

Draw and color the correct shapes missing in each pattern.

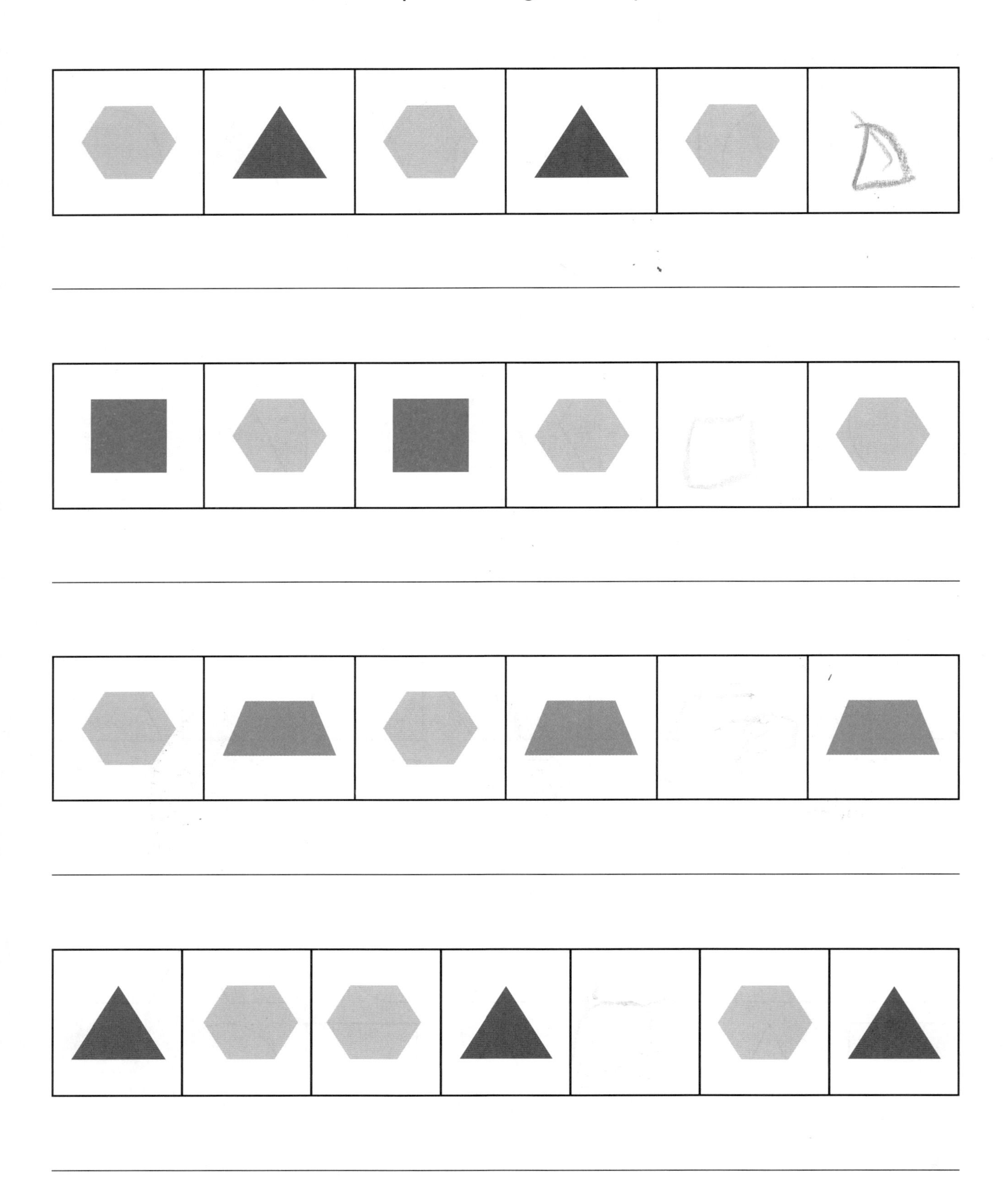

Using this page: Have students continue each pattern by drawing and coloring the correct shape in the box.
Concept: Identifying patterns.

Exercise 8

Color the shapes of the same size.

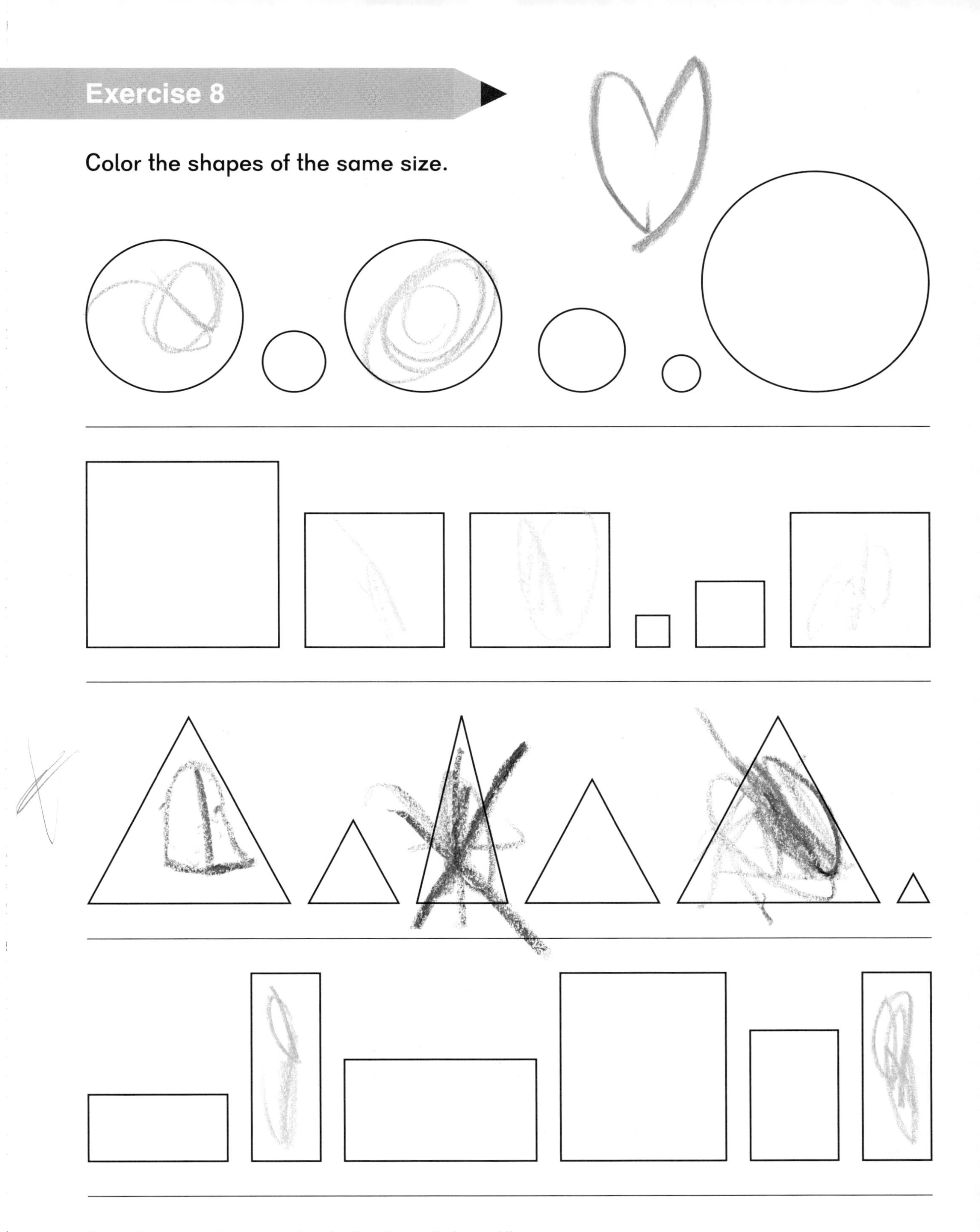

Using this page: Have students color the shapes that are of the same size.
Concept: Identifying sizes and shapes.

Draw the shape that completes the pattern.

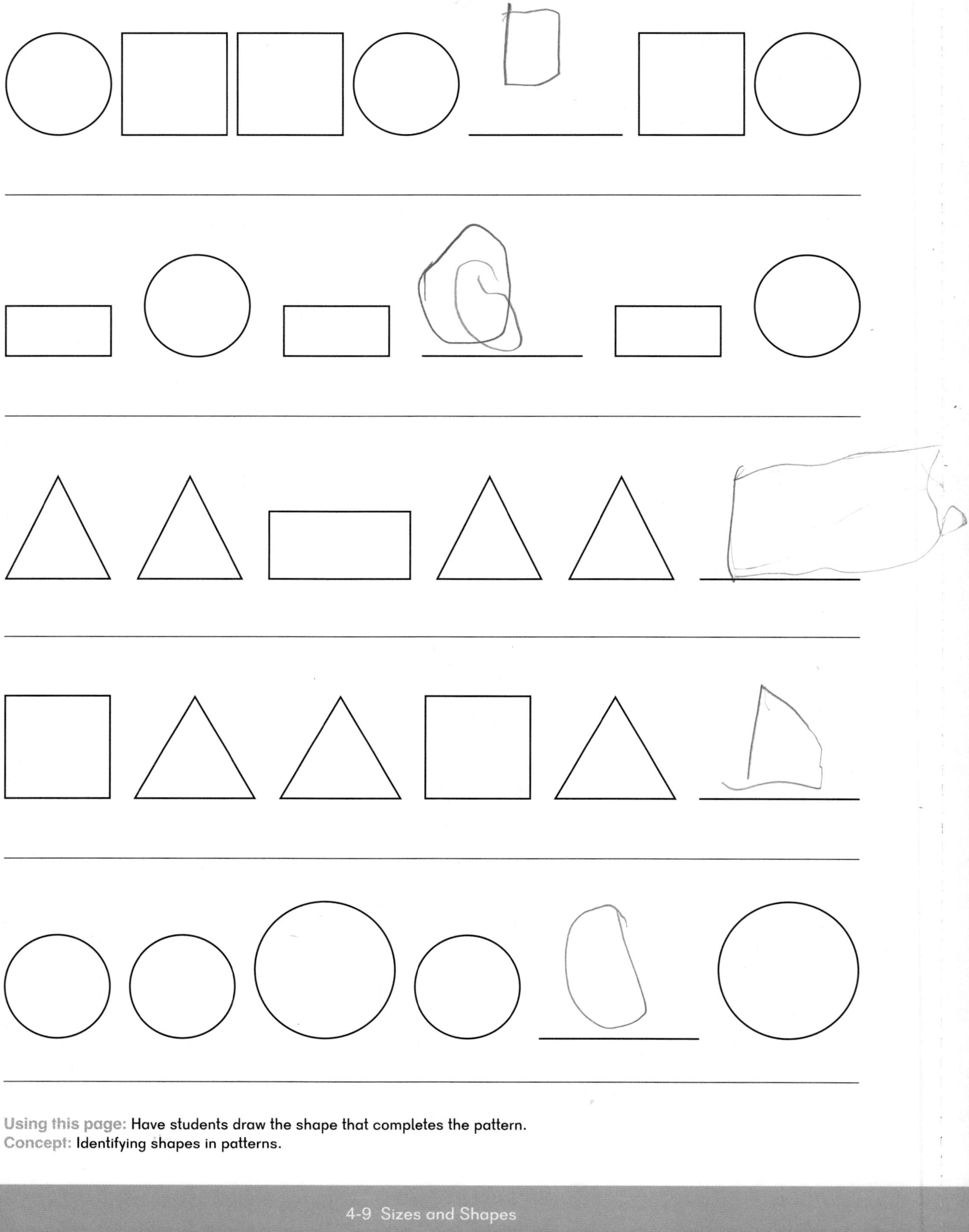

Using this page: Have students draw the shape that completes the pattern.
Concept: Identifying shapes in patterns.

Exercise 9

Copy the picture and write the number.

Before using this page: Distribute pattern blocks comprising of squares, triangles, hexagons, and trapezoids to students.
Using this page: Have students copy the picture to reproduce the dog with pattern blocks, then count and record the number of blocks used for each shape in the box.
Concept: Identifying and positioning shapes with partial outlines.

Create your own shapes picture!

Before using this page: Pre-cut the shapes on page 149.
Using this page: Have students create their own picture using the shape cut-outs.
Concept: Creating with shapes.

Exercise 10

Count and color the graph.
Write the numbers.

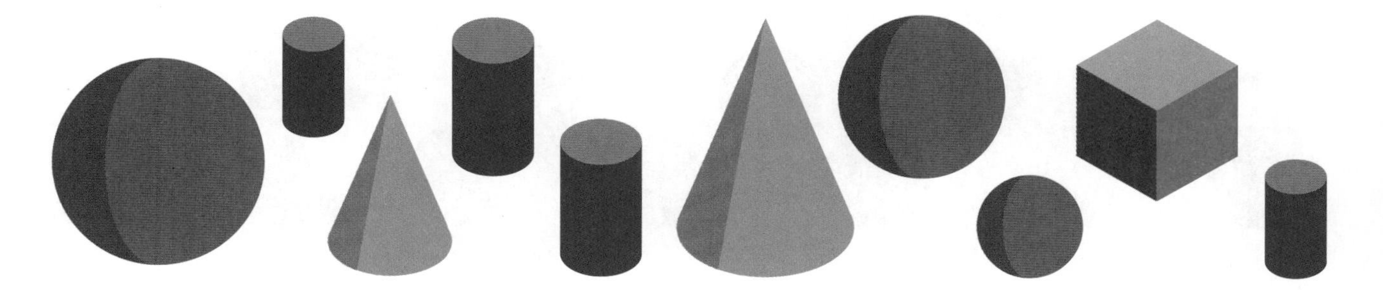

Favorite Solids			

Using this page: Have students count each solid and color in the respective boxes to complete the graph, then write its numeral in the box at the bottom.
Concept: Creating picture graphs.

Count and color the graph.
Write the numbers.

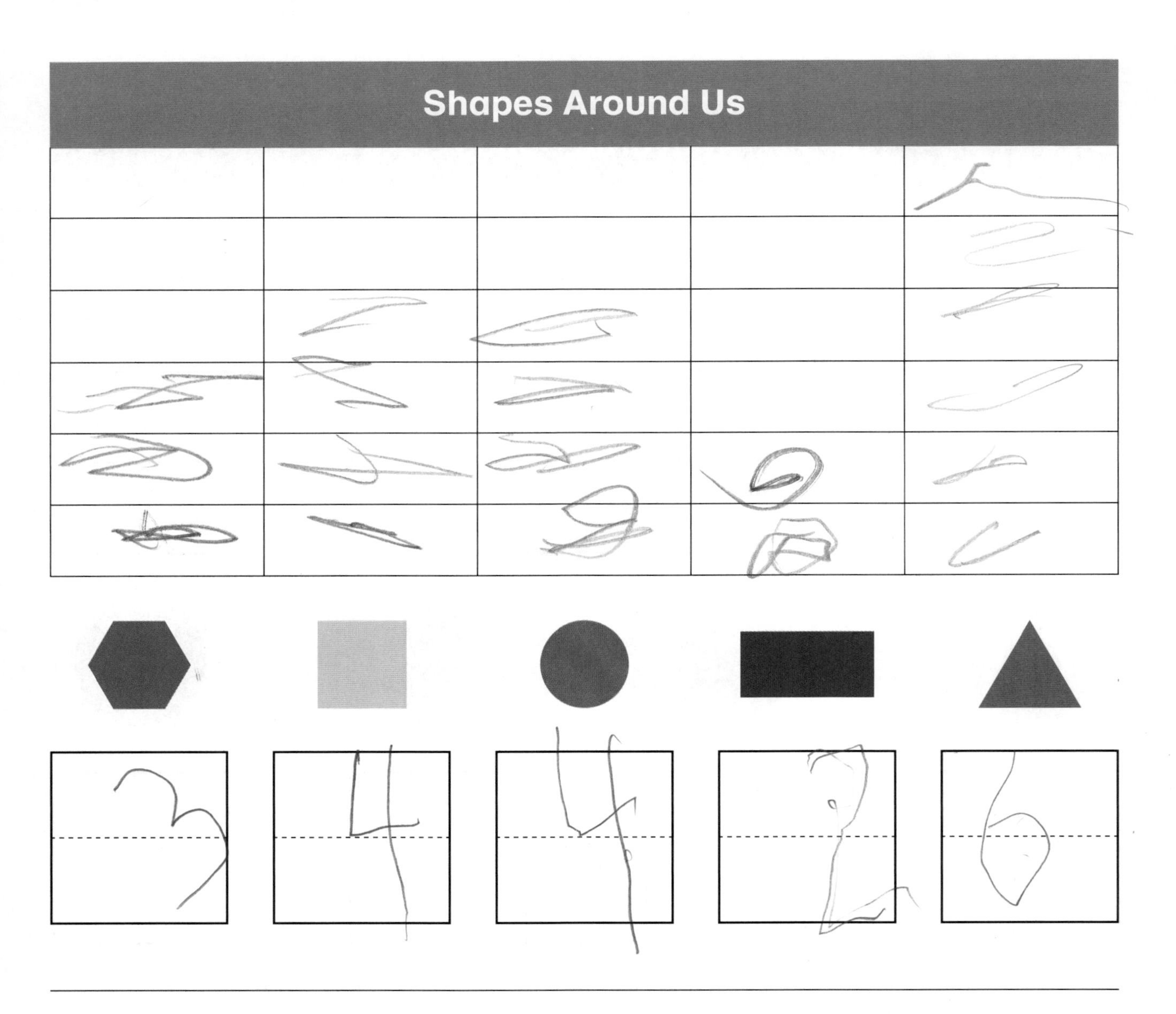

Using this page: Have students count each shape and color in the respective boxes to complete the graph, then write its numeral in the box at the bottom.
Concept: Creating picture graphs.

Exercise 11

Color according to the Color Key.
Count and write the number.

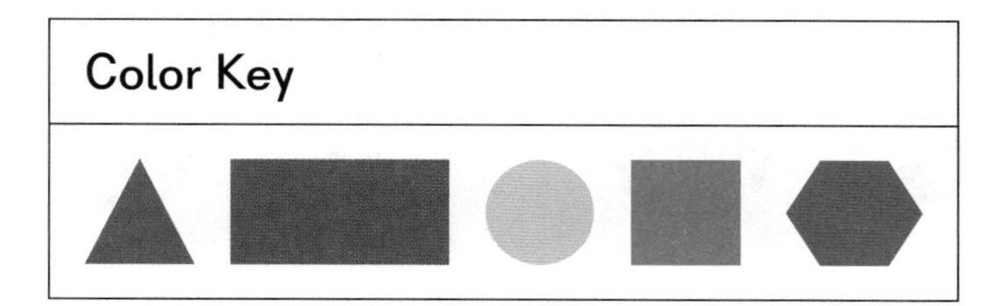

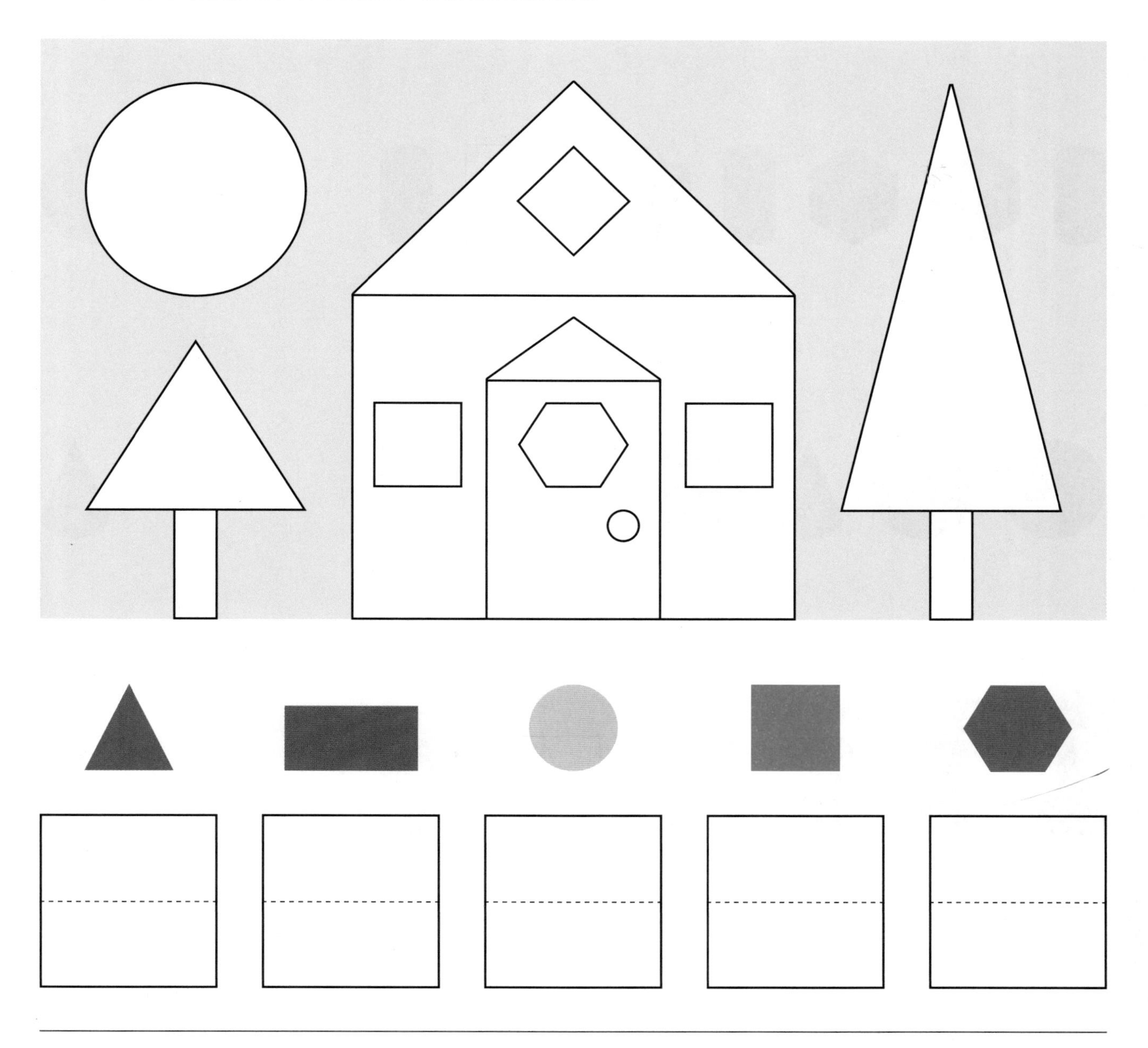

Using this page: Have students refer to the color key and color the shapes, then count each shape and write its numeral in the box.

Match to complete the pattern.

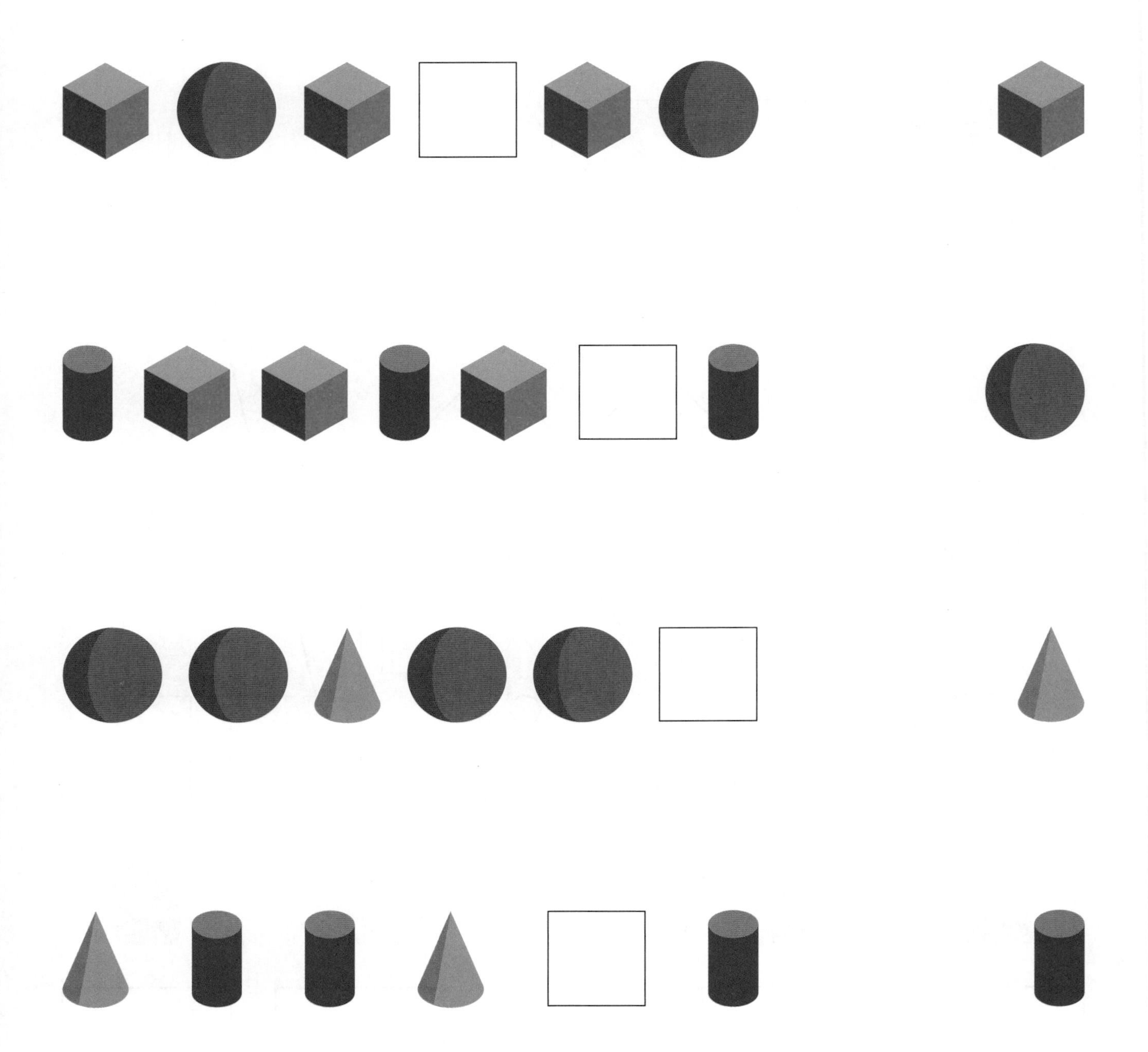

Using this page: Have students find what's missing in each pattern and match to the correct solid.

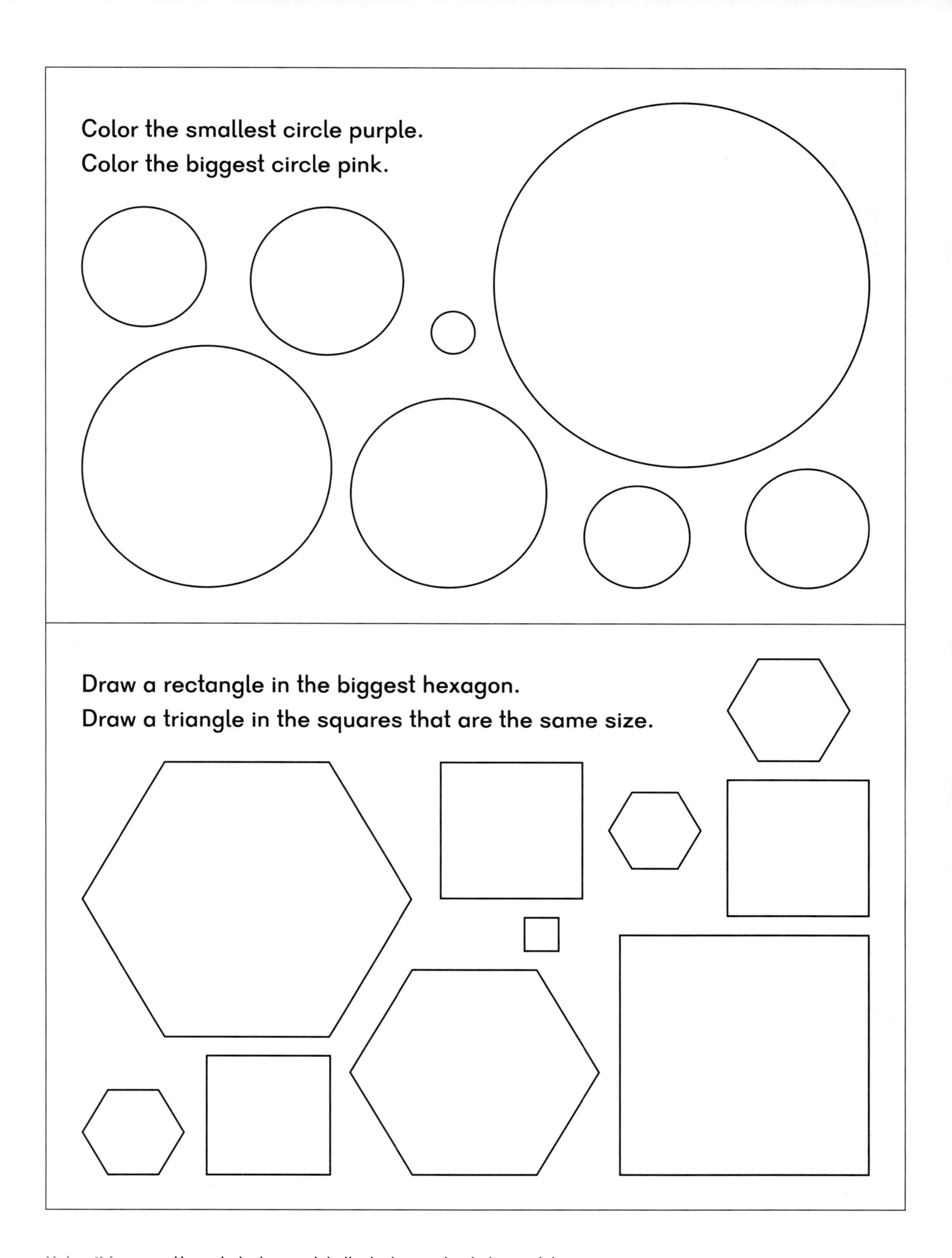

Using this page: Have students complete the task as instructed in each box.

Find the hidden shapes and color according to the Color Key.

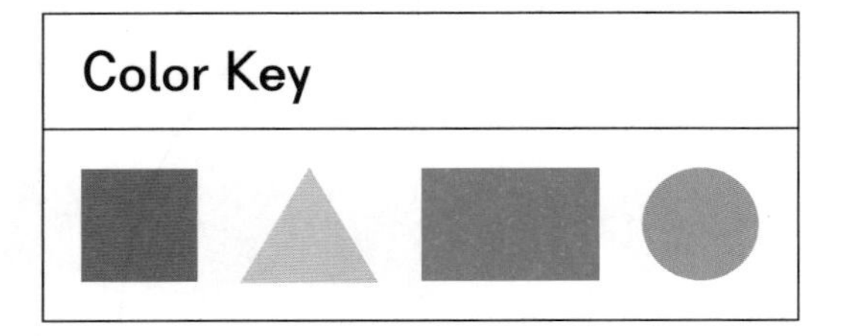

Using this page: Have students identify the shapes and color.

Chapter 5 Compare Height, Length, Weight, and Capacity

Exercise 1

Circle the tallest thing.
Cross out the shortest thing.

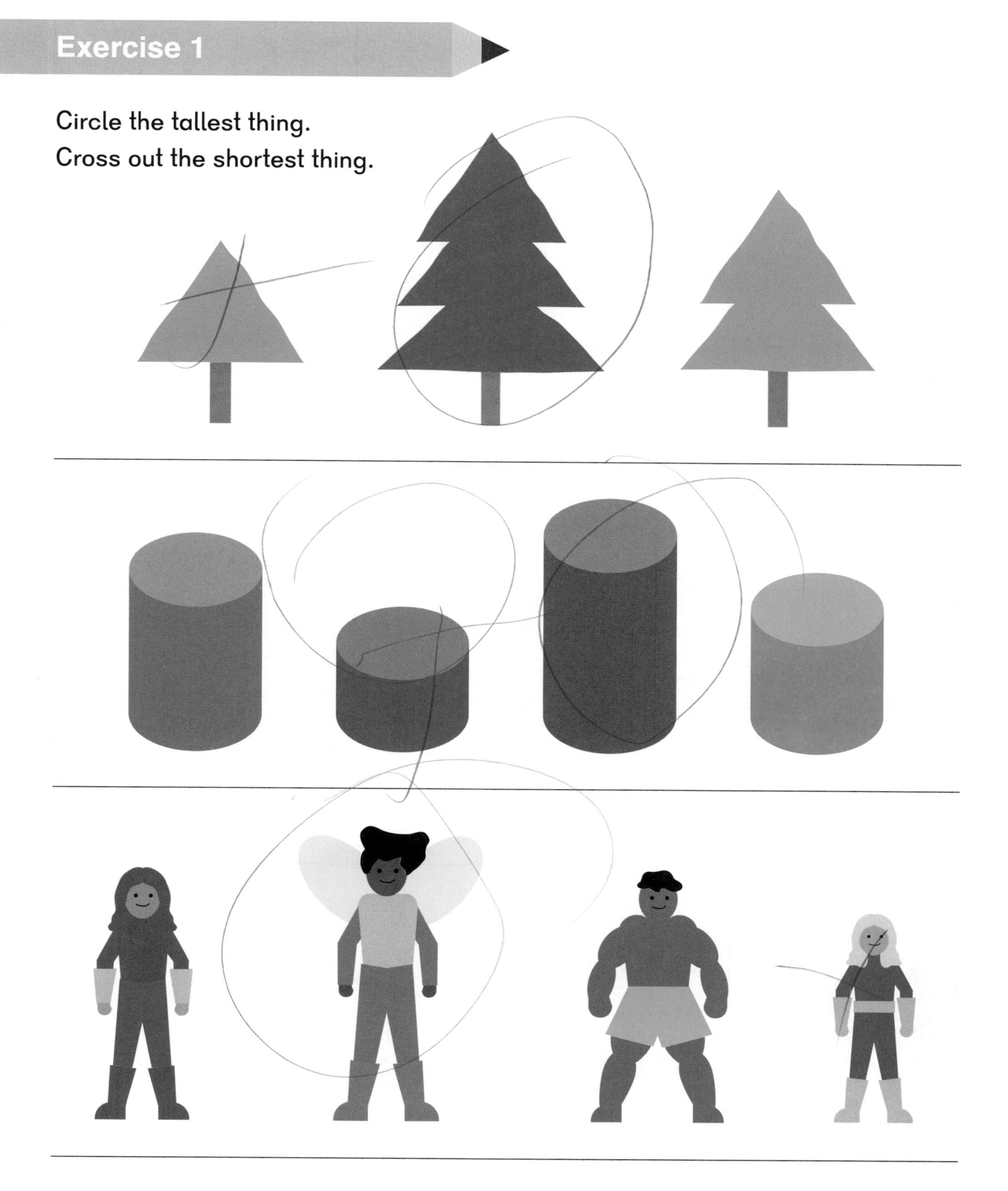

Using this page: Have students compare the objects in each row, then circle the tallest and cross out the shortest.
Concept: Comparison of height.

Order the giraffes from tall to tallest.

Order the toadstools from short to shortest.

Before using this page: Guide students in cutting the pictures on page 151.
Using this page: Have students order the giraffes from tall to tallest and paste in space provided. Next, have them order the toadstools from short to shortest and paste.
Concept: Comparing and ordering of height.

Exercise 2

Color the longest thing yellow.
Color the shortest thing pink.

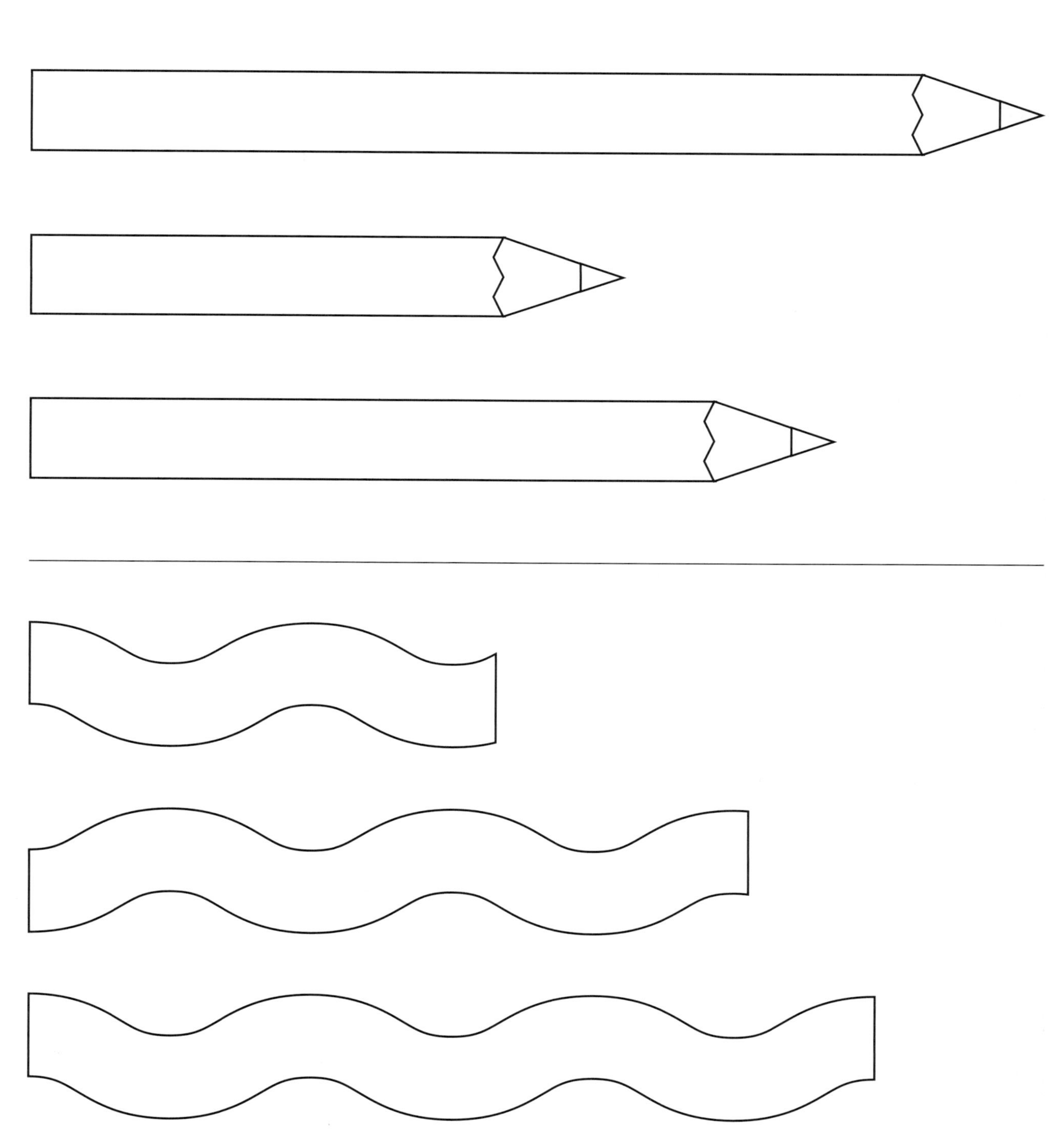

Using this page: Have students compare each group of objects, then follow the directions and color.
Concept: Comparison of length.

Order the rulers from long to longest.

Before using this page: Pre-cut the rulers on page 153.
Using this page: Have students arrange the cut-out rulers in order from long to longest and paste.
Concept: Ordering length from long to longest.

Exercise 3

Mark the height of the ladders.

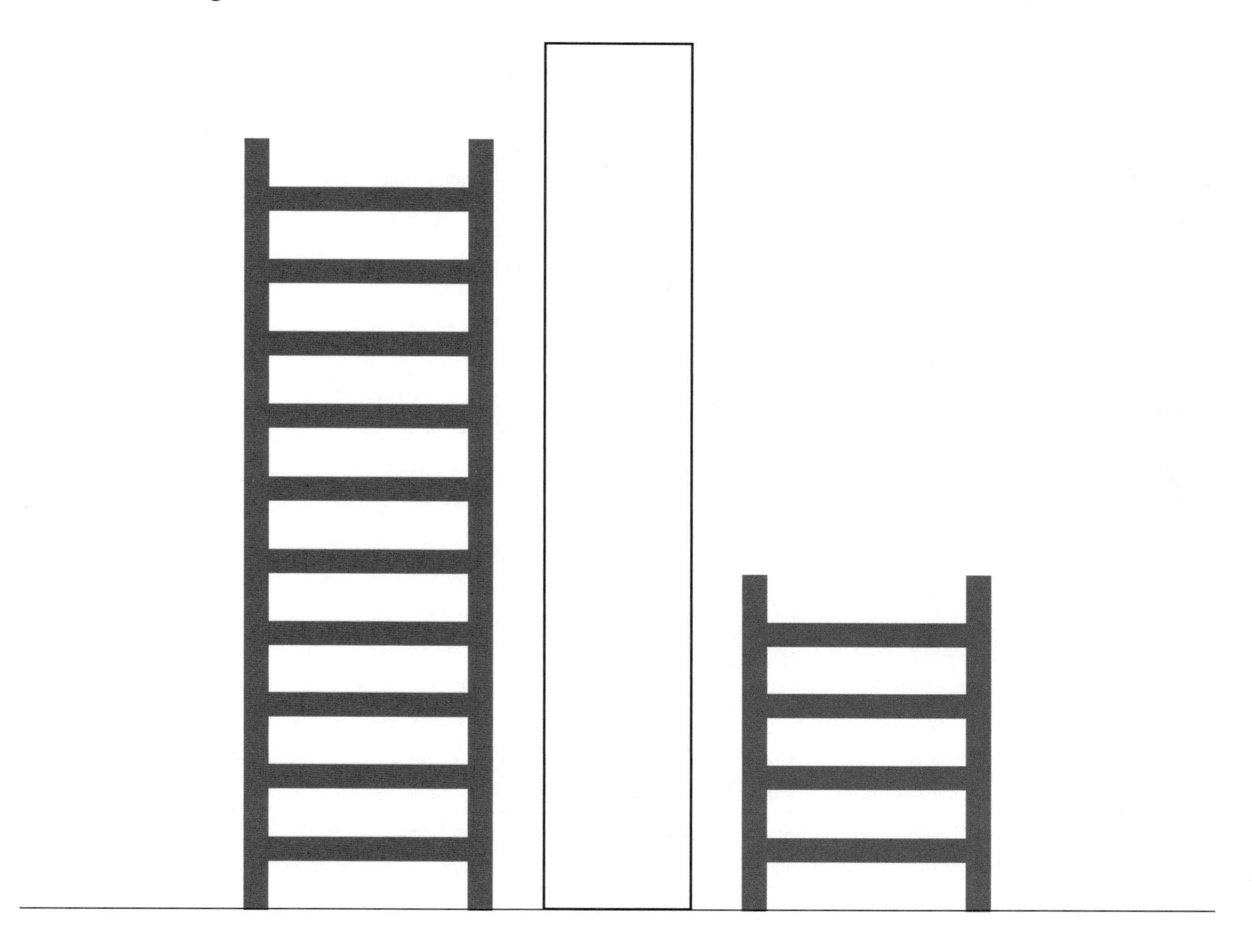

Circle the taller ladder below.

This  is taller.

Using this page: Have students mark the height of each ladder on the paper tape between them with a ruler and pencil, then circle the picture of the taller ladder at the bottom of the page.
Concept: Comparing height using a rod.

Mark the length of the animals.

Circle the longest and the shortest animal below.

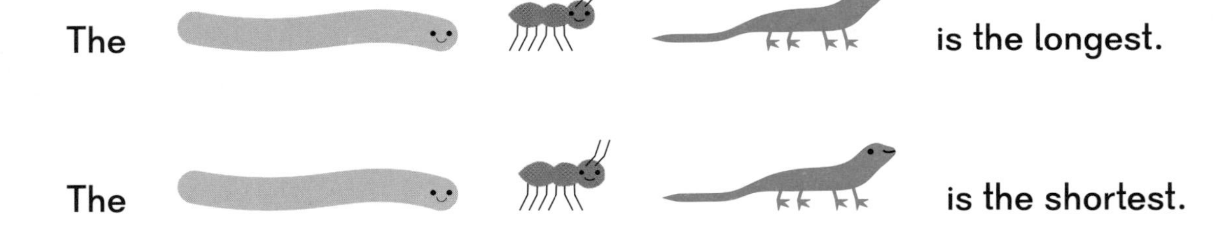

The [worm] [ant] [lizard] is the longest.

The [worm] [ant] [lizard] is the shortest.

Using this page: Have students mark the length of each animal on the paper tape using the same color as each animal.
Concept: Comparing length using a paper tape.

Exercise 4

Color and write the number.

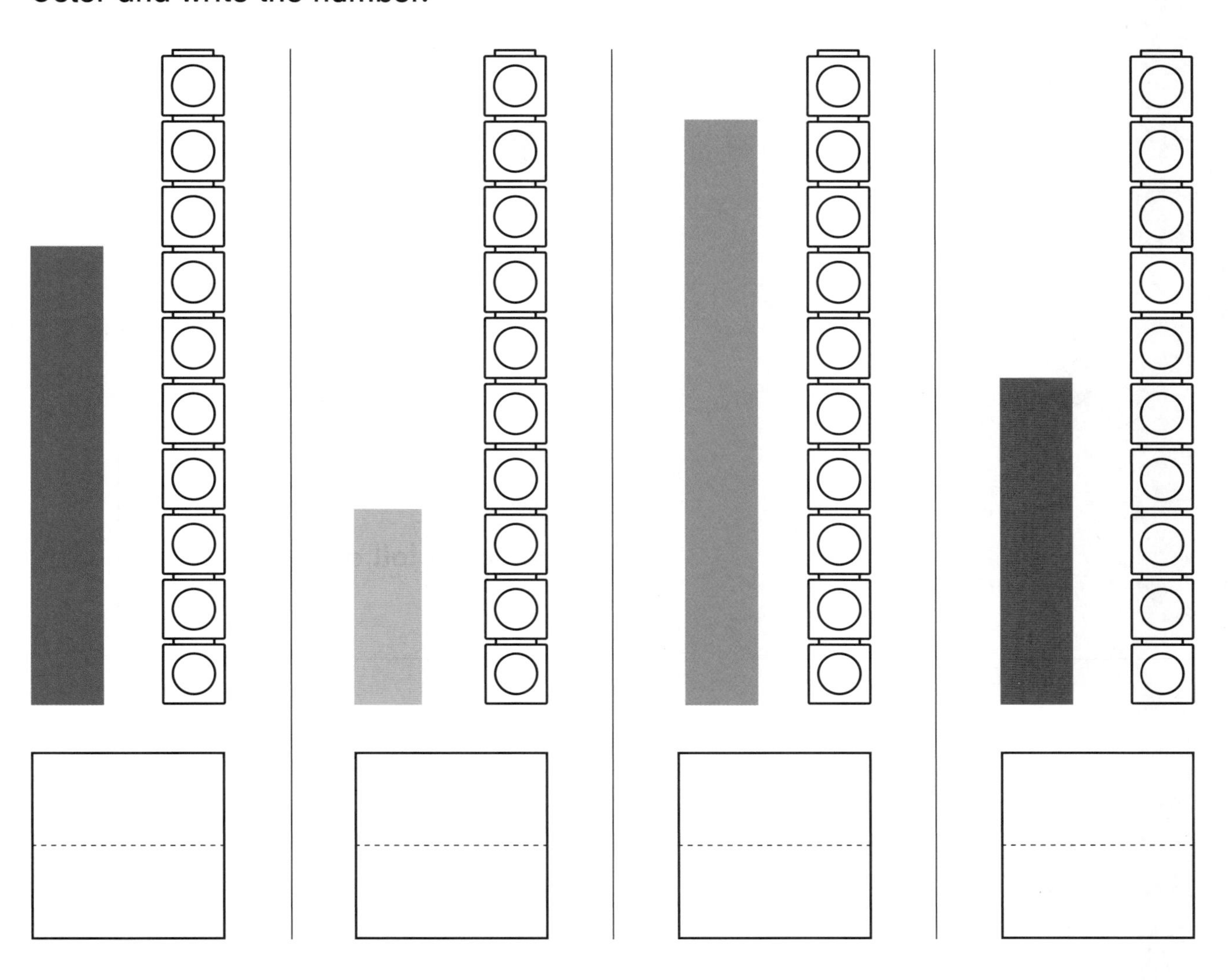

Circle the tallest and shortest rod below.

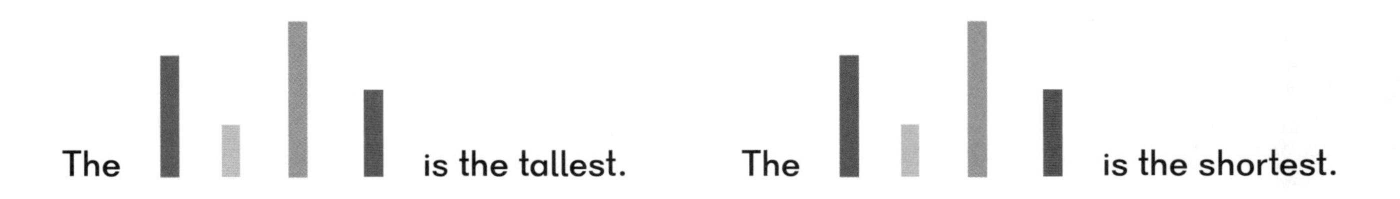

Using this page: Have students color the cubes that measure the height of the rods, then write the last number counted in the box. Then have them circle the tallest and shortest rod at the bottom of the page.
Concept: Comparing height using cubes.

Count and write the number.

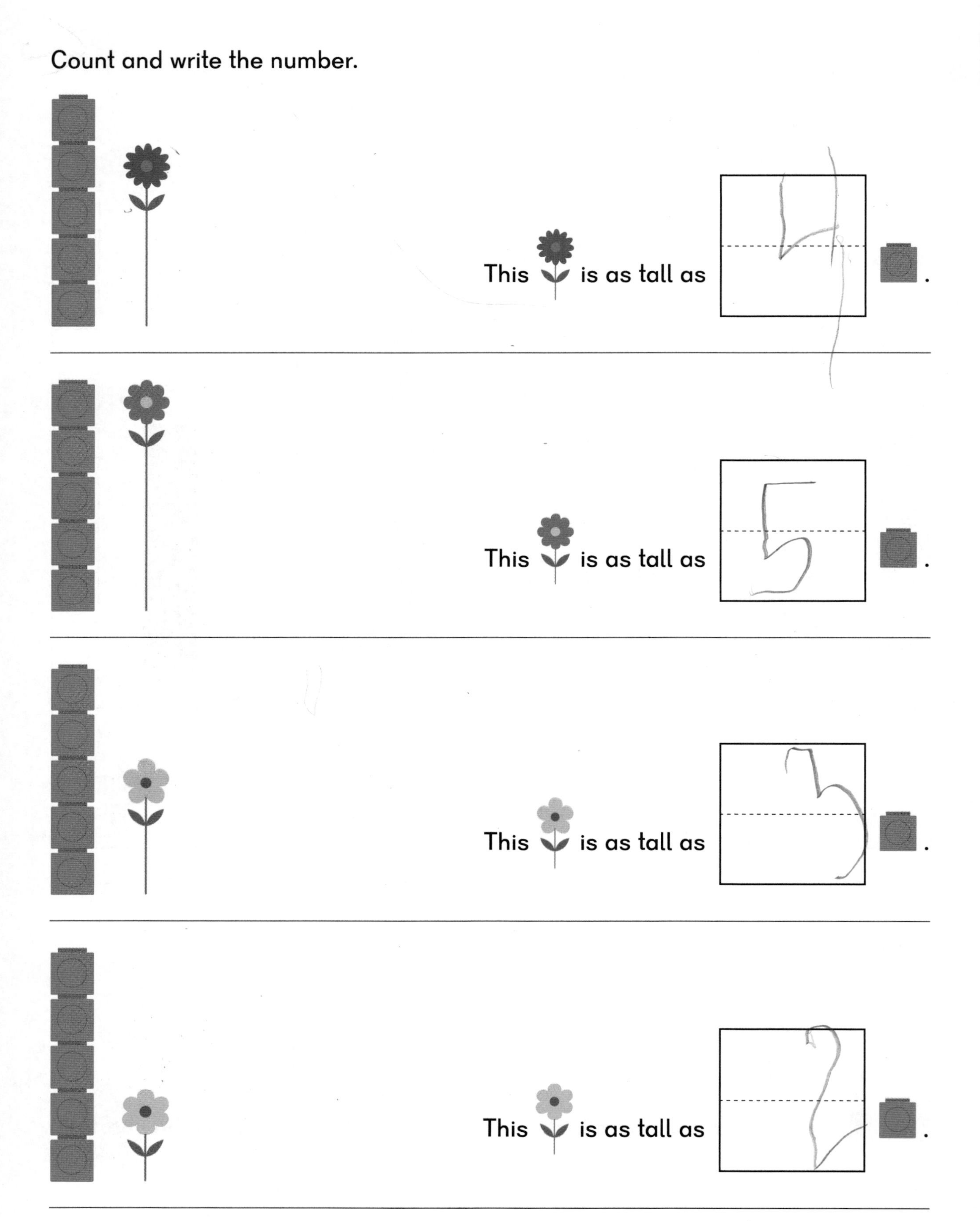

Using this page: Have students count the cubes to measure the height of the flowers, then write the numeral in the box.
Concept: Measuring height using cubes.

Count and write the number.
Color the longest pencil blue.
Color the shortest pencil orange.

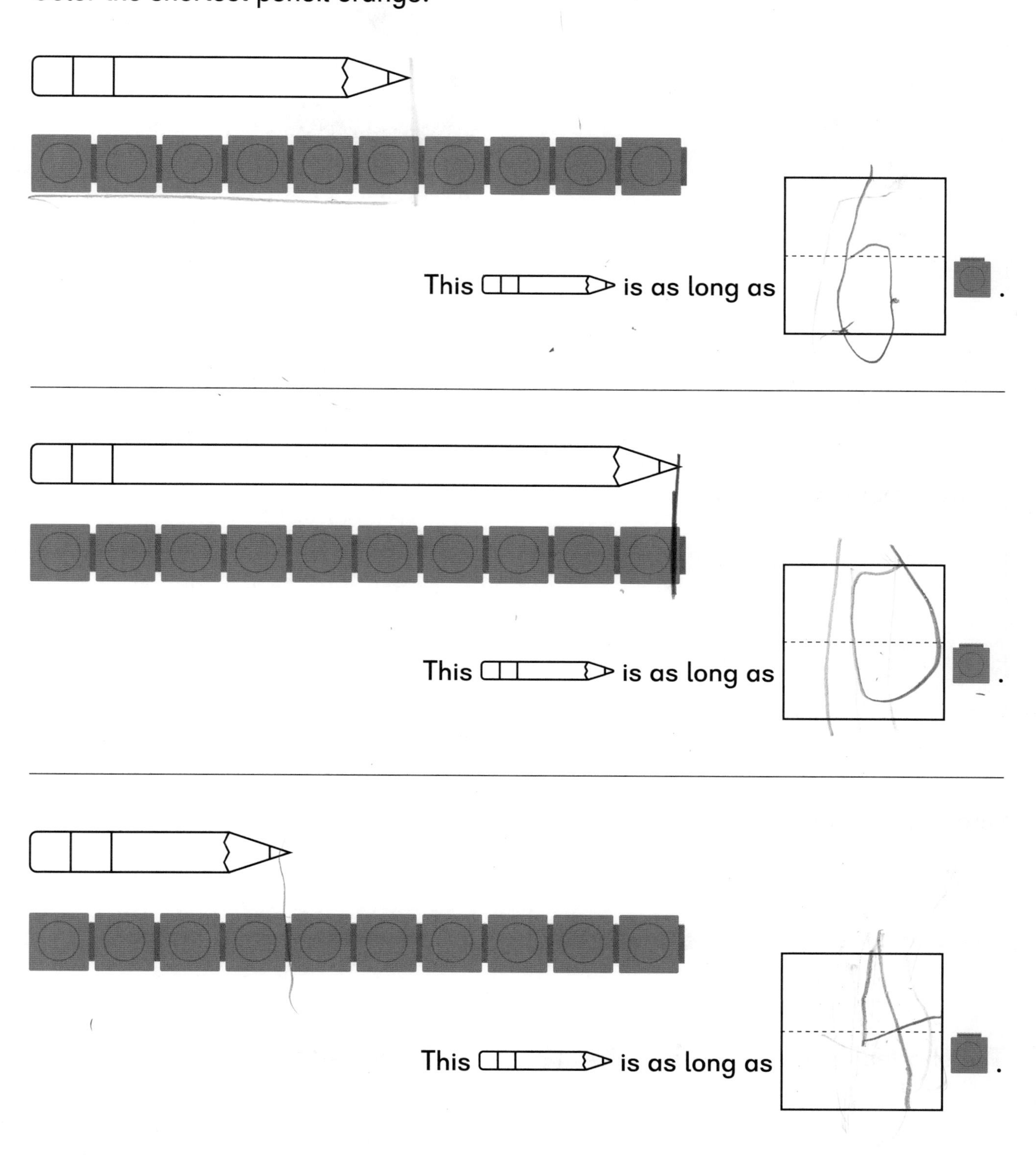

Using this page: Have students count the cubes that measure the length of each pencil and write the numeral in the box. Then have them color the longest and shortest pencils as specified.
Concept: Measuring length using cubes.

Count and write the number.

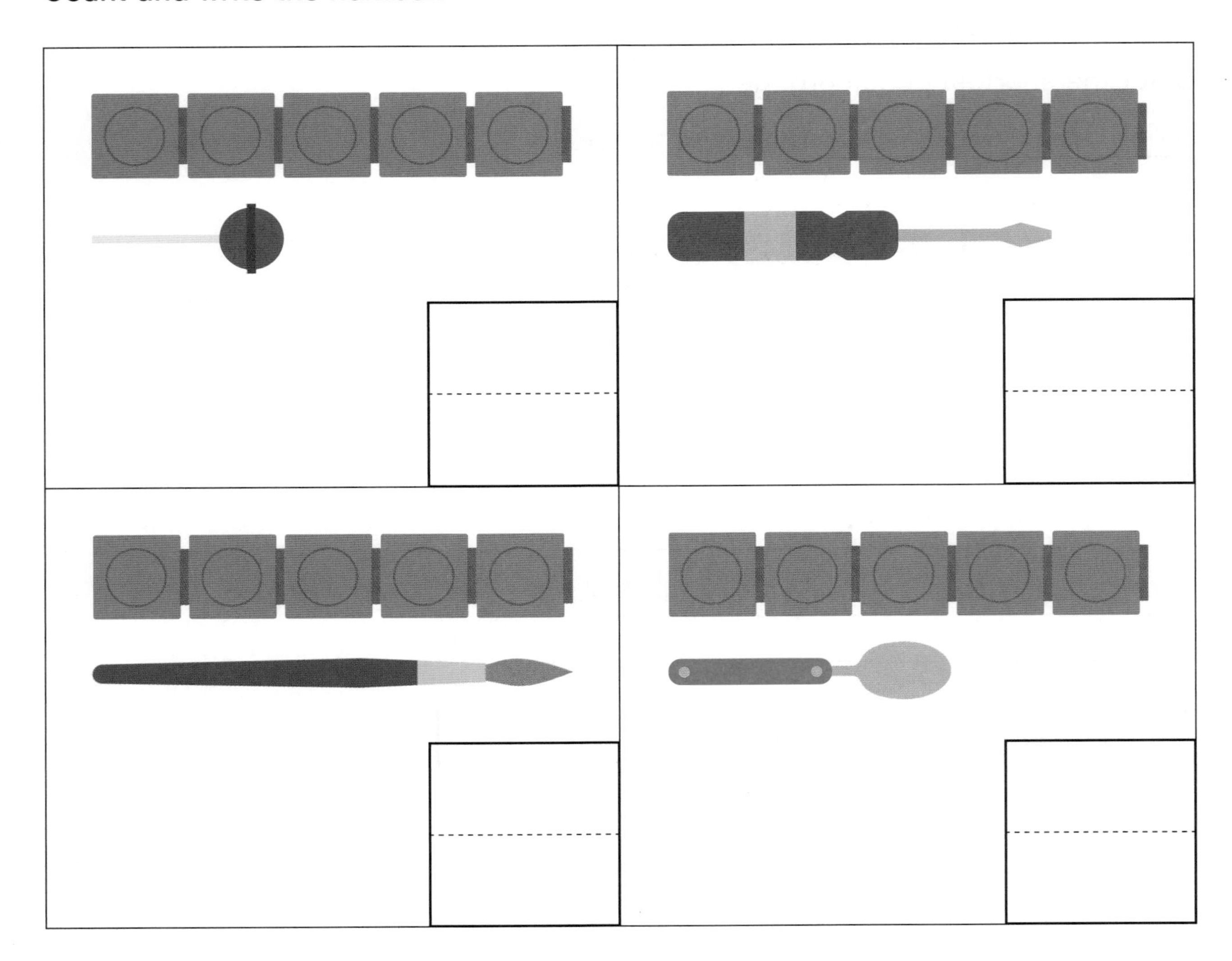

Circle the longest and shortest thing.

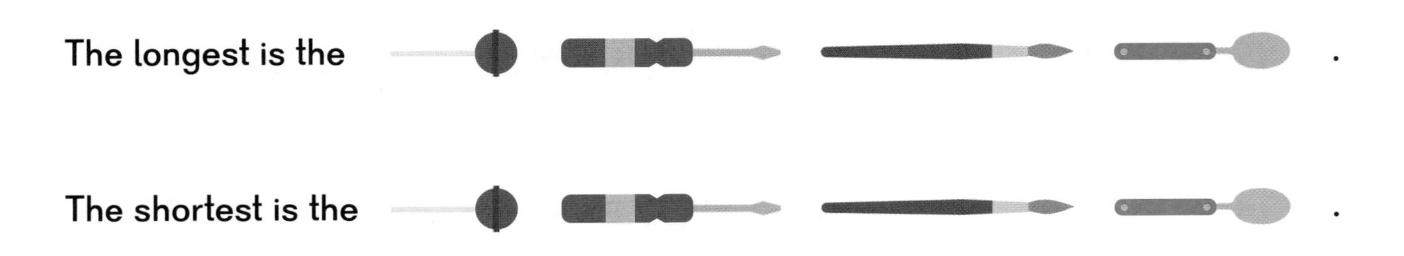
The longest is the .

The shortest is the .

Using this page: Have students count the cubes that measure the length of the objects and write the numeral in the box. Then have them compare their length and circle the correct picture at the bottom.
Concept: Comparing length using cubes.

Exercise 5

Circle the heavier thing.

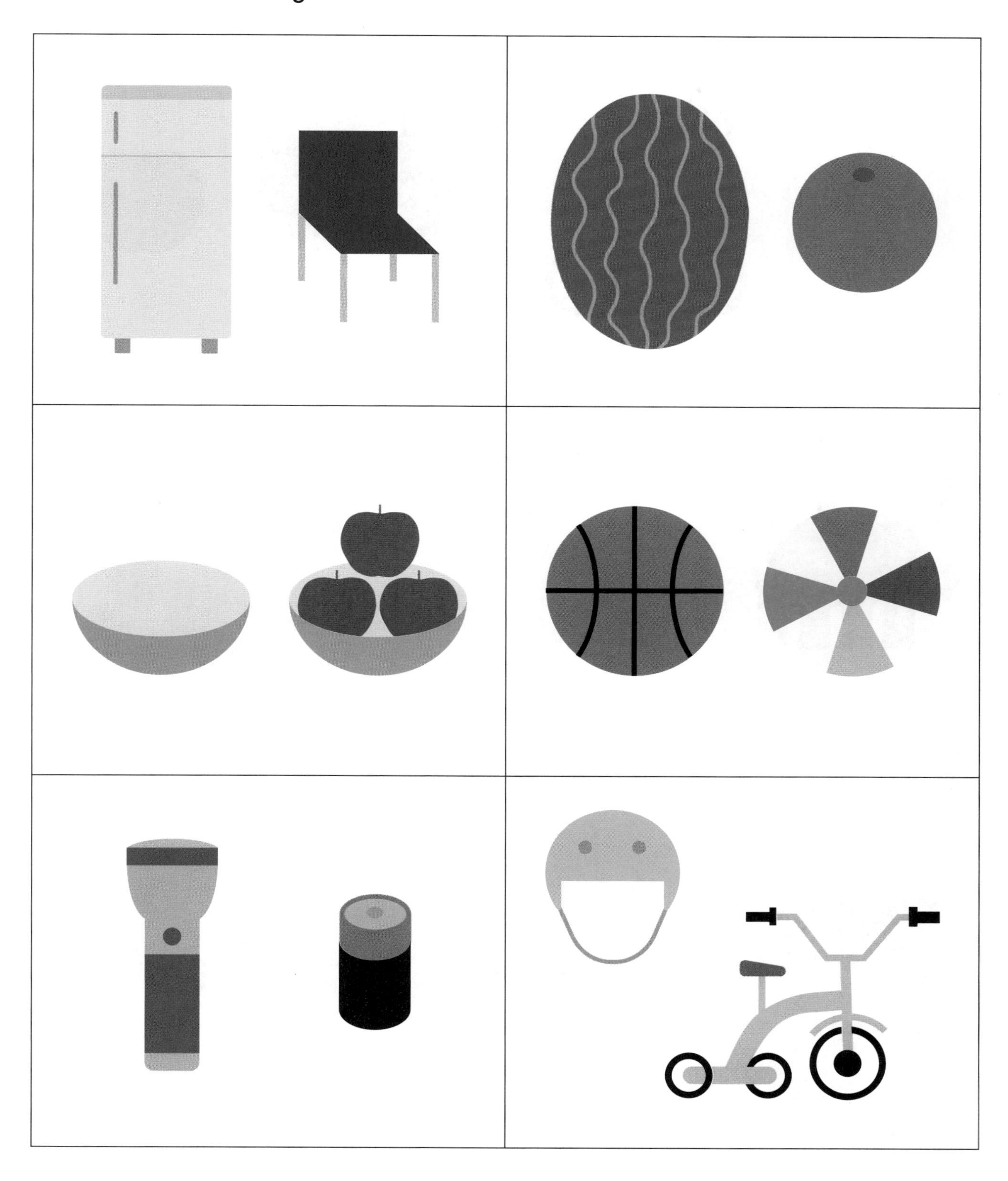

Using this page: Have students compare and circle the heavier object in each box.
Concept: Comparing weight of things around us.

Circle the lighter one.

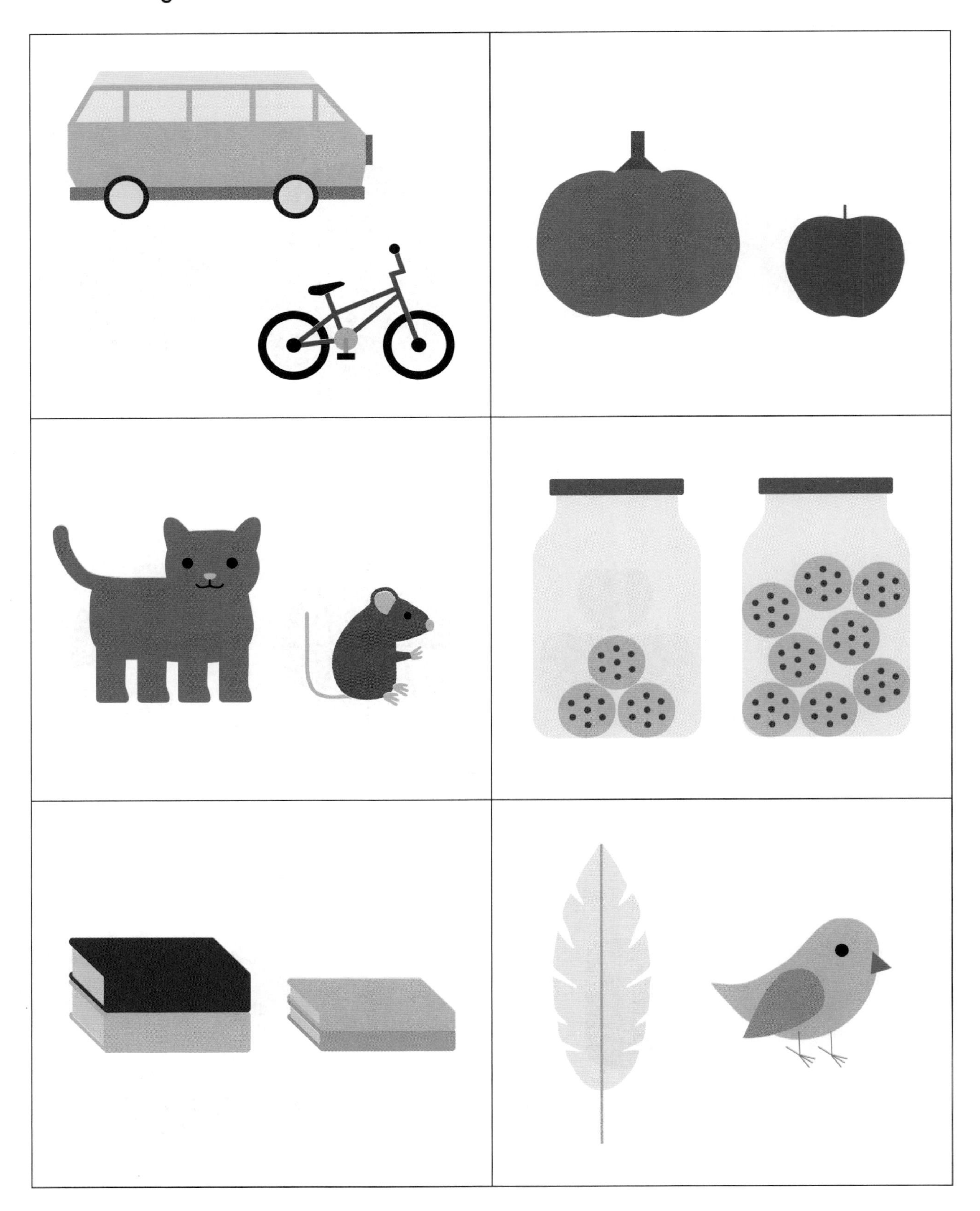

Using this page: Have students compare and circle the lighter object in each box.
Concept: Comparing weight of things around us.

Exercise 6

Circle the heavier thing.

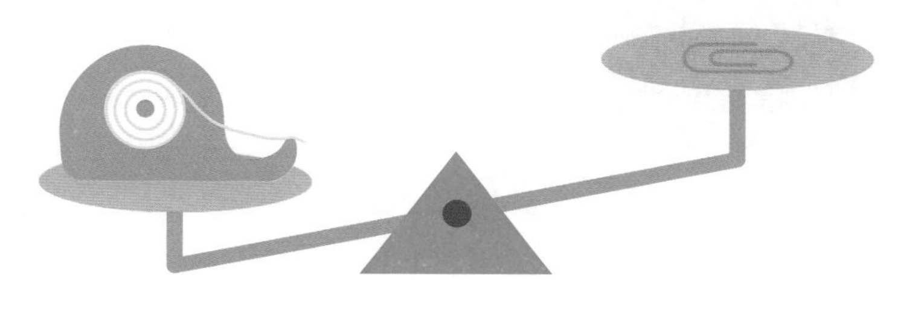

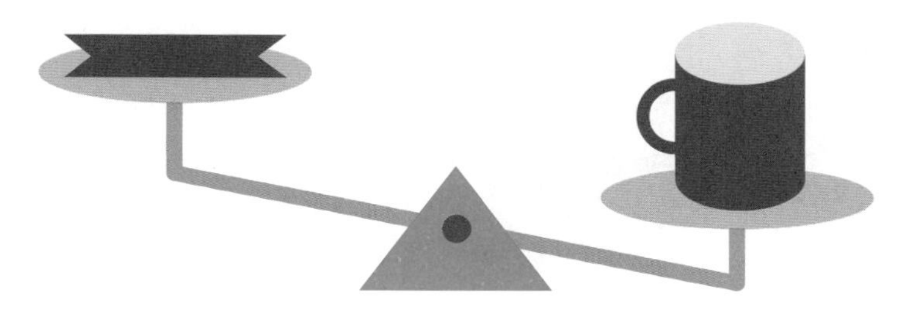

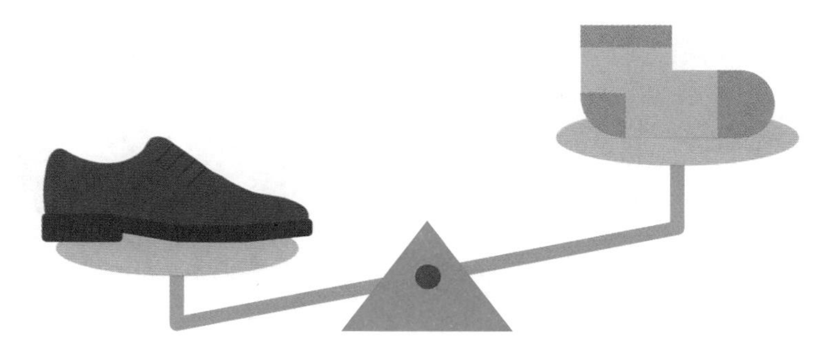

Using this page: Have students circle the heavier object on each balance scale.
Concept: Comparing weight on a balance scale.

Circle the lighter thing.

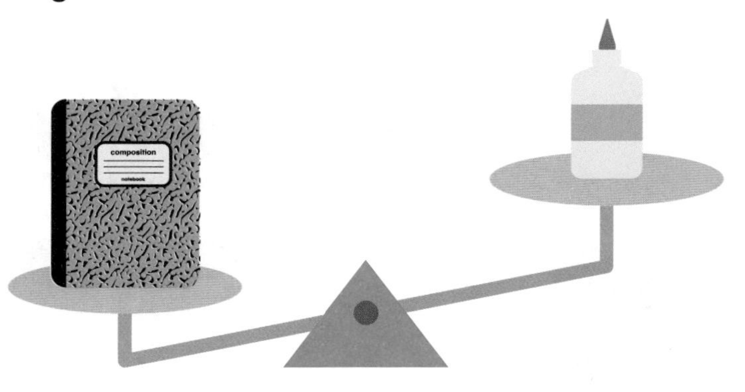

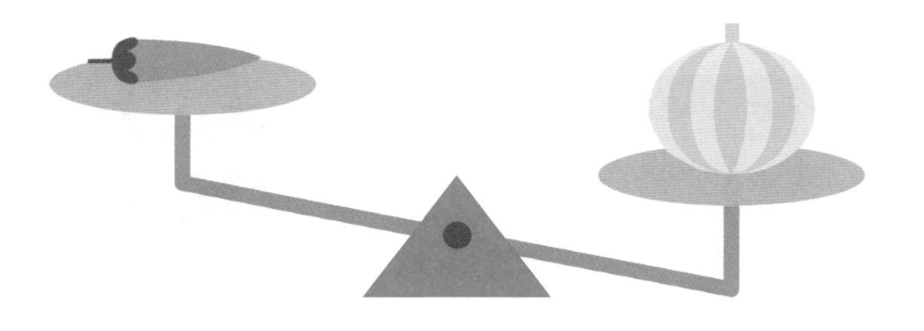

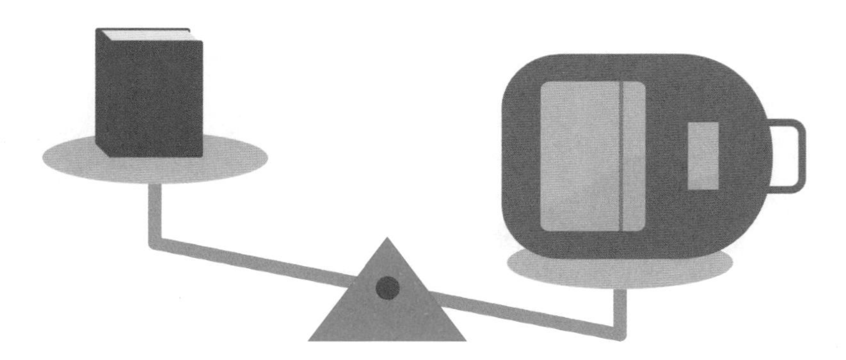

Using this page: Have students circle the lighter object on each balance scale.
Concept: Comparing weight on a balance scale.

Exercise 7

Count and write the number.

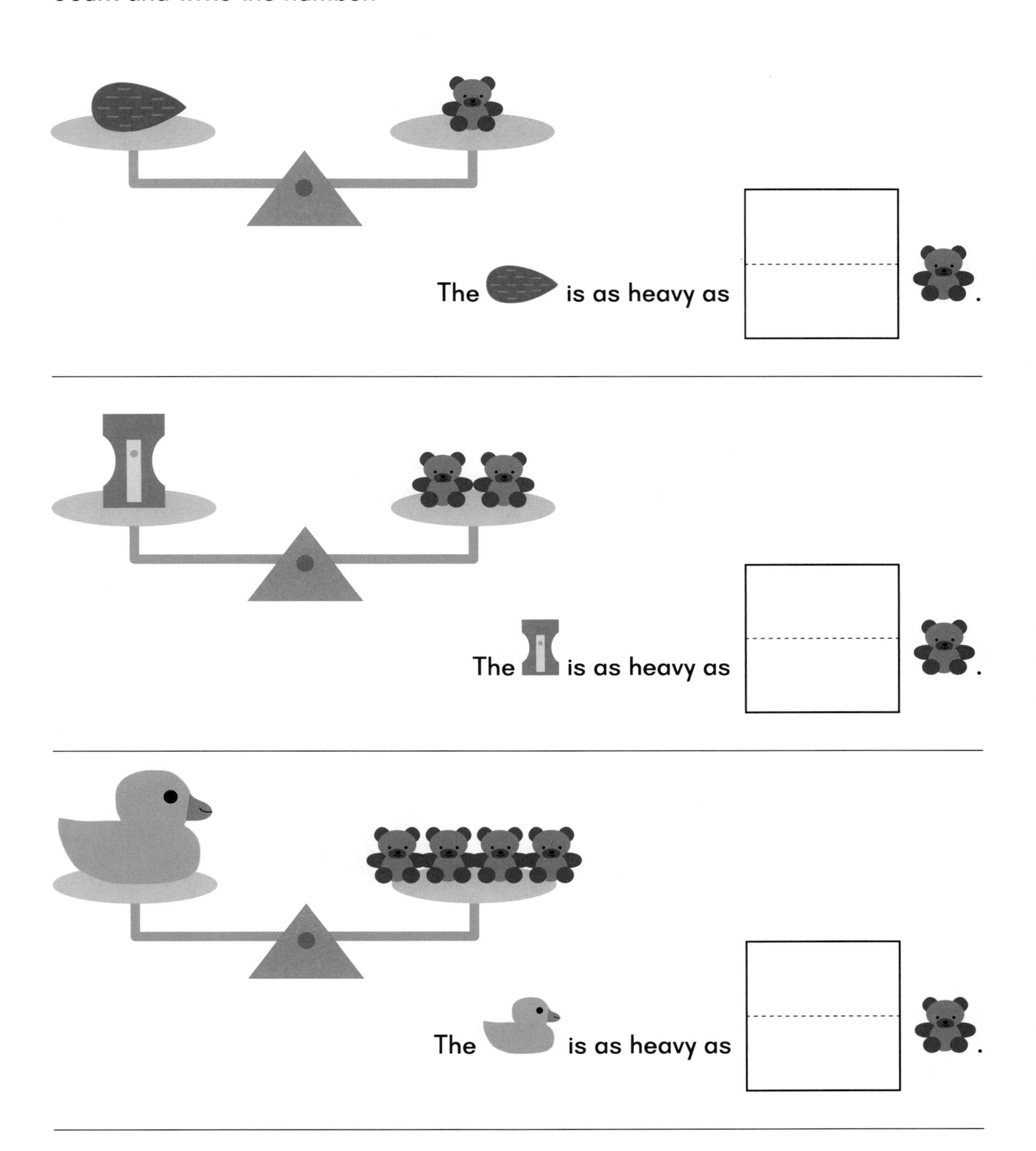

Using this page: Have students find how many teddy bear counters weigh as much as each object and write the numeral in the box.
Concept: Measuring weight using teddy bear counters.

Count and write the number.

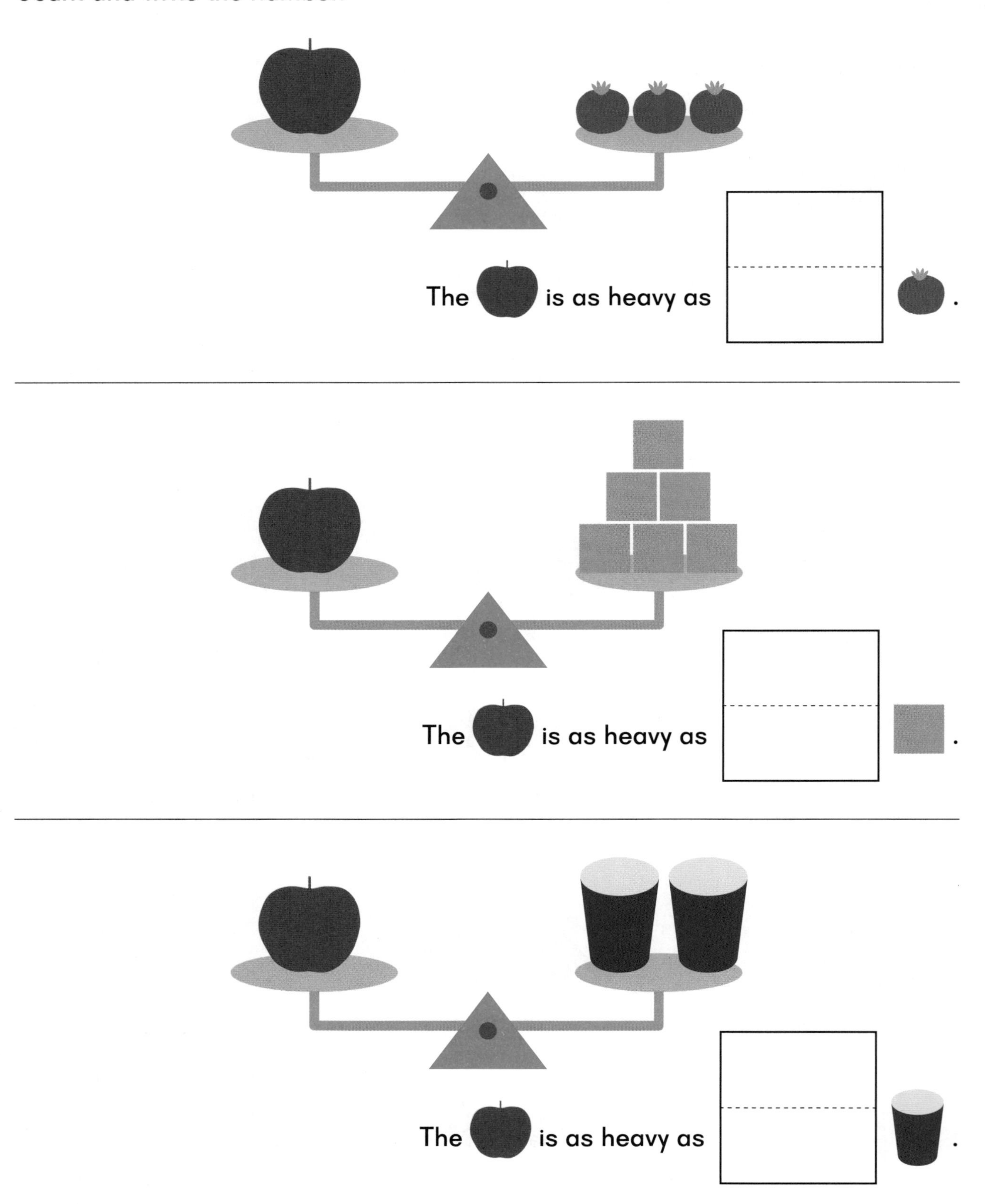

Using this page: Have students find how many of the given objects weigh as much as an apple and write the numeral in the box.
Concept: Measuring weight using non-standard units.

Exercise 8

Circle the container that can hold more.

Using this page: Have students circle the container that holds more liquid than the other.
Concept: Comparing capacity.

Circle the container that holds less.

Using this page: Have students circle the container that holds less liquid than the other.
Concept: Comparing capacity.

Exercise 9

Count and write the number.

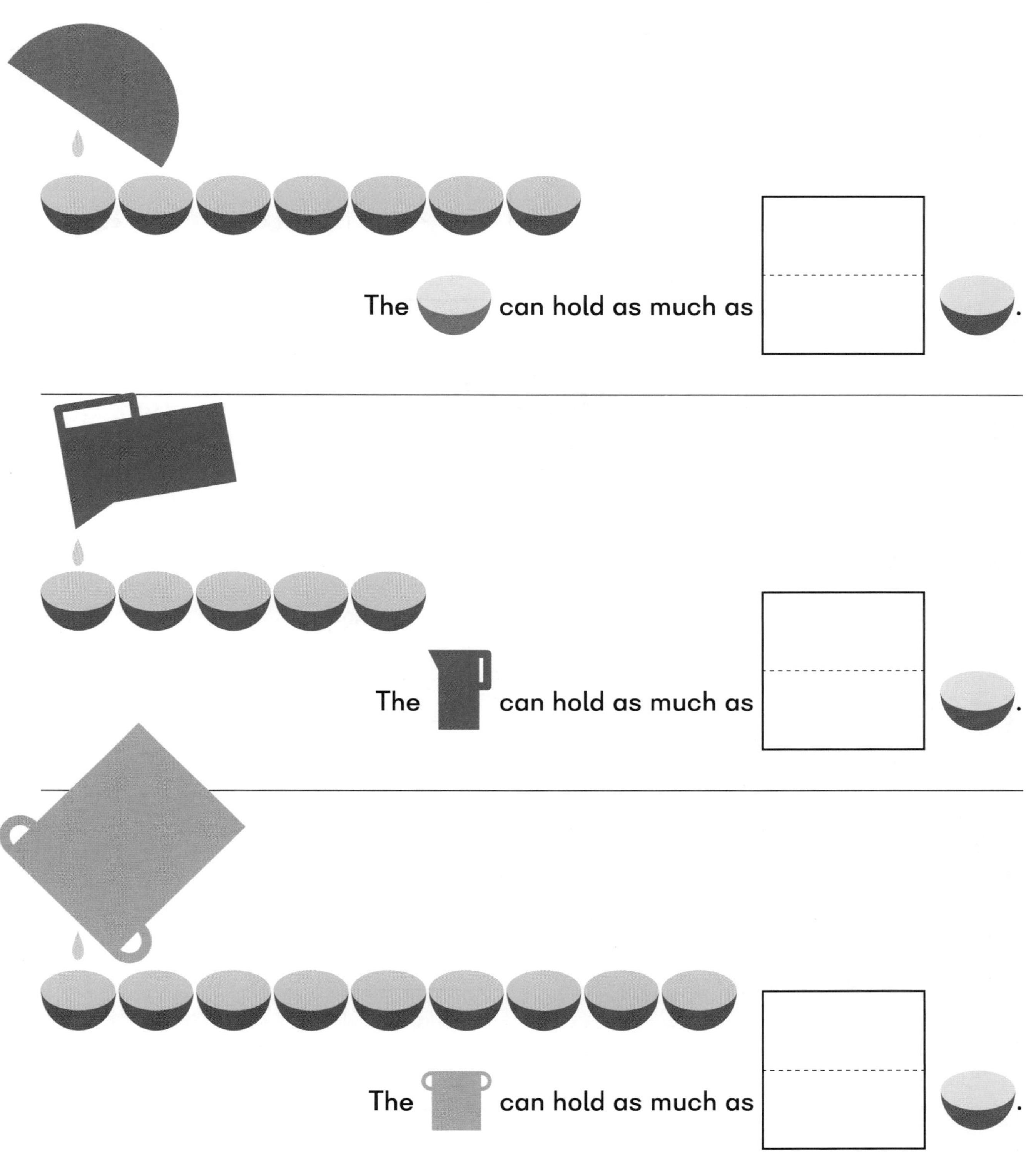

Using this page: Have students count the number of bowls and write the numeral in the box.
Concept: Measuring capacity using bowls.

Count and write the number.

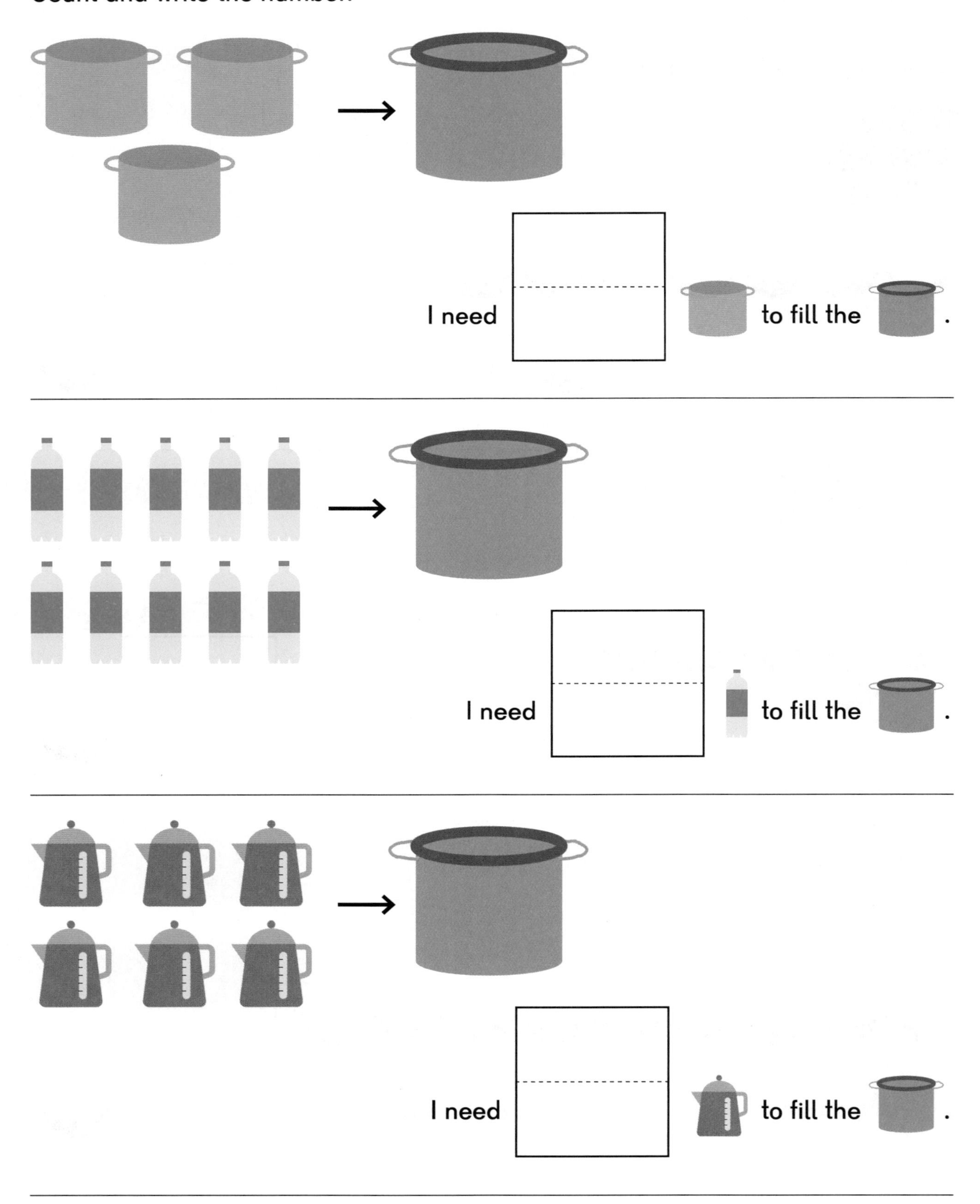

Using this page: Have students count the number of objects and write the numeral in the box.
Concept: Measuring capacity using non-standard units.

Exercise 10

Count and write the number.
Circle the shortest bottle.
Cross out the tallest bottle.

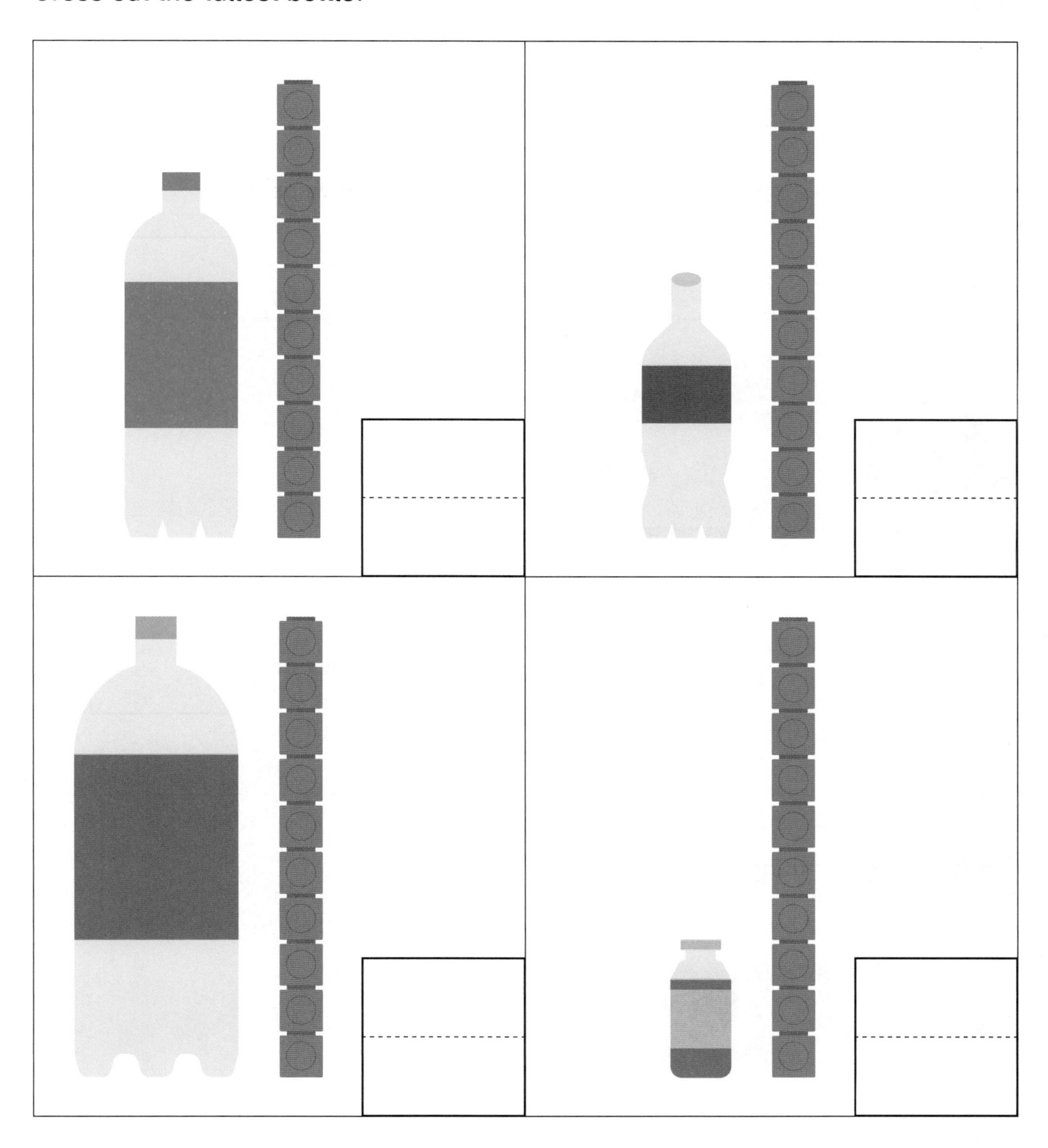

Using this page: Have students count the cubes that measure the height of the bottles and write the numeral in the box. Then have them circle the shortest bottle and cross out the tallest bottle.

Count and write the number.

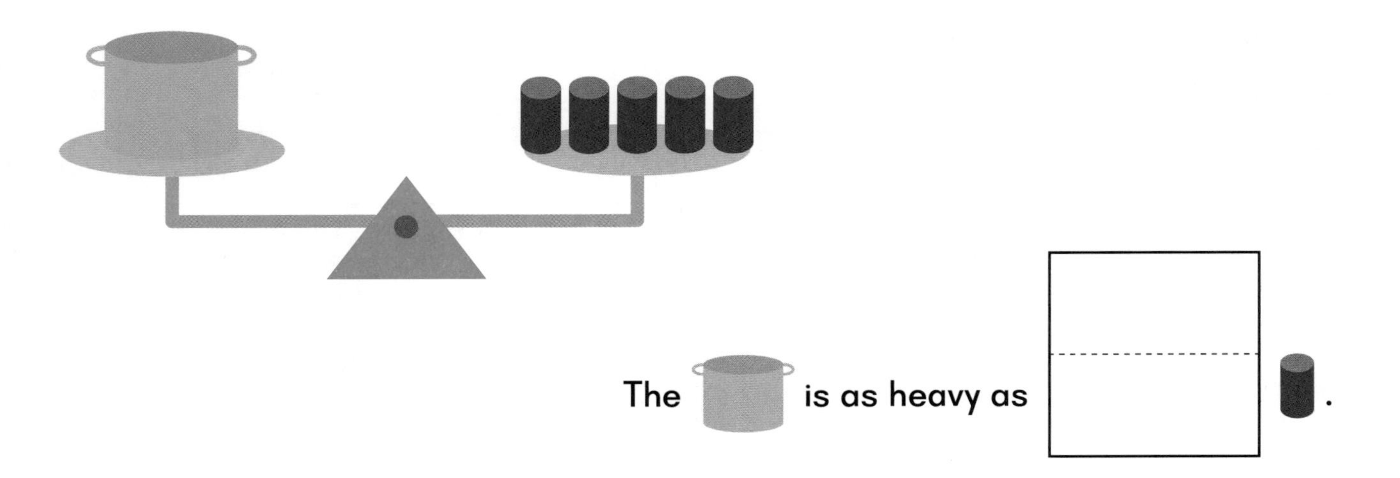

The [pot] is as heavy as [ ] [cylinder].

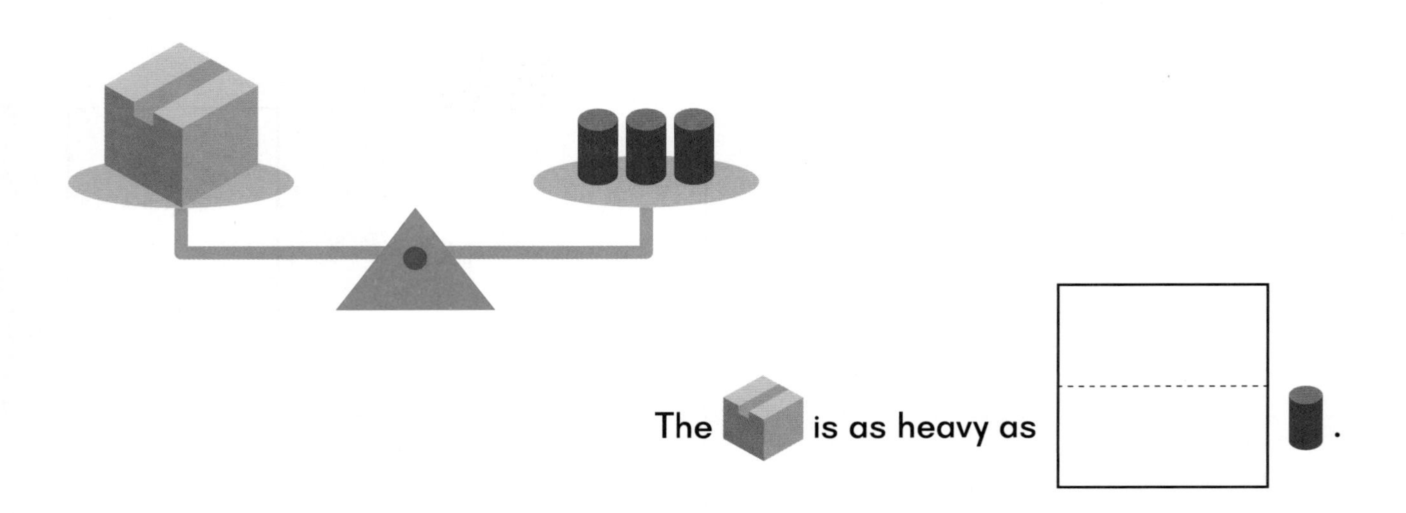

The [box] is as heavy as [ ] [cylinder].

Circle the lighter thing.

The [pot] [box] is lighter.

Using this page: Have students find how many cylinders weigh as much as the objects and write the numeral in the box. Then have them circle the correct picture below.

Chapter 6 Comparing Numbers Within 10

Exercise 1

Circle the 2 groups that have the same number.

Using this page: Have students match the two sets to find out if they have the same number, then circle the equal sets.
Concept: Identifying equal sets.

Color the 2 groups with the same number.

Using this page: Have students match the two sets to find out if they have the same number, then color the equal sets.
Concept: Identifying equal sets.

Circle the group that has more.

Using this page: Have students compare the two sets to find which set has more, then circle the set that has more.
Concept: Identifying the set that has more.

Color the group that has more.

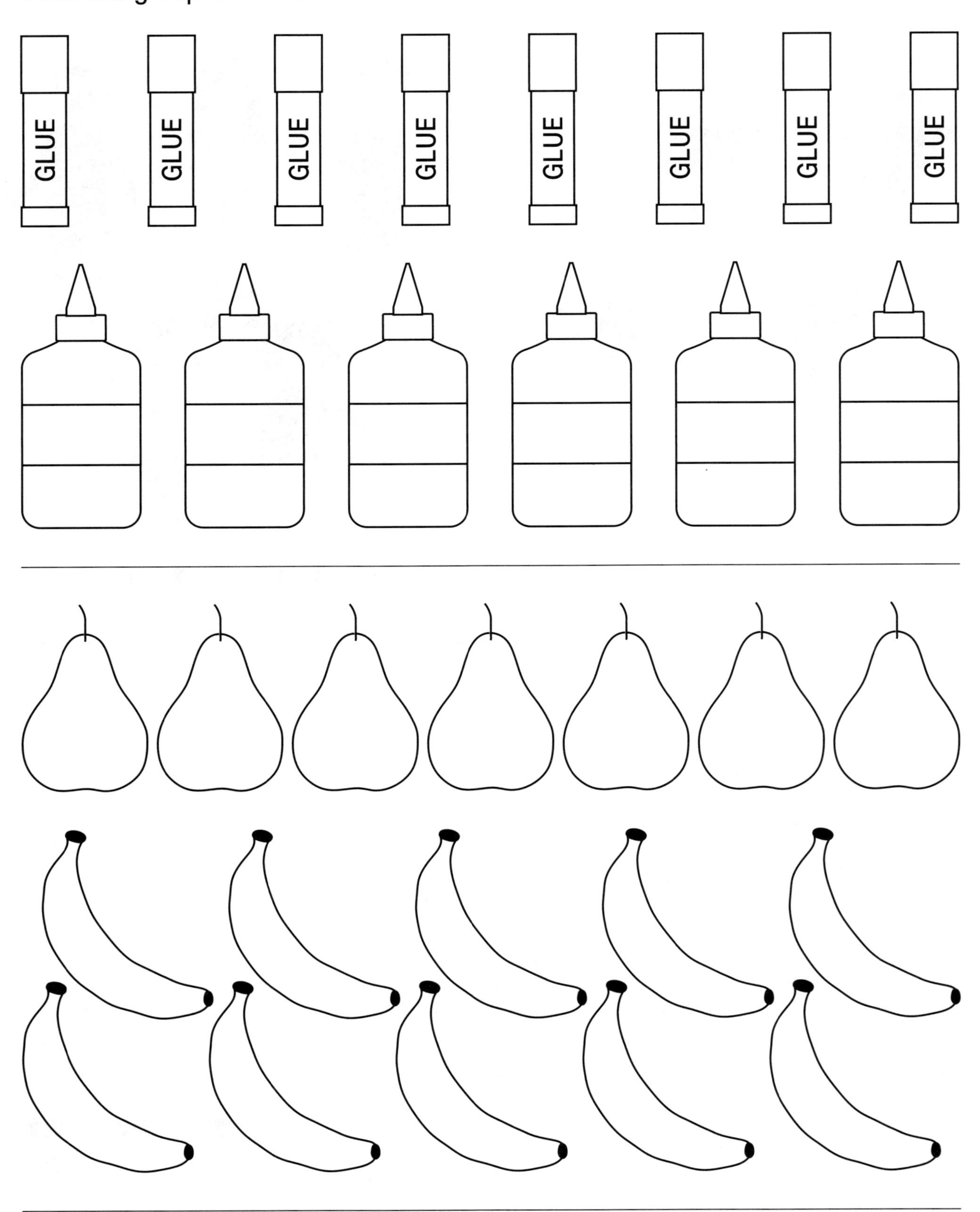

Using this page: Have students compare the two sets of objects to find out which one has more, then color that set.
Concept: Identifying the set that has more.

Exercise 2

Circle the group that has more.

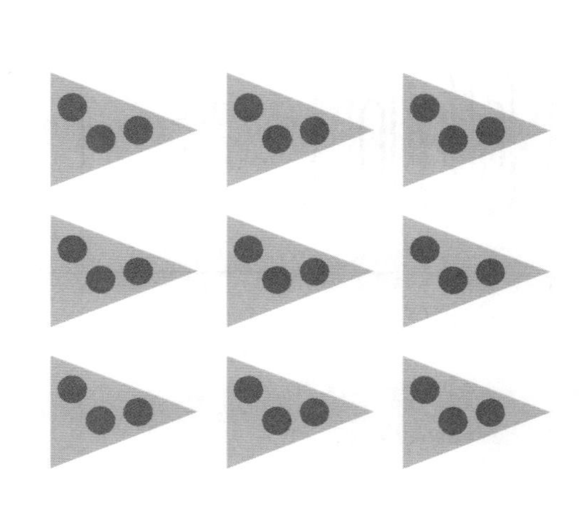

Using this page: Have students circle the set that has more in each row.
Concept: Identifying the set that has more.

Write the number and circle the group that has more.

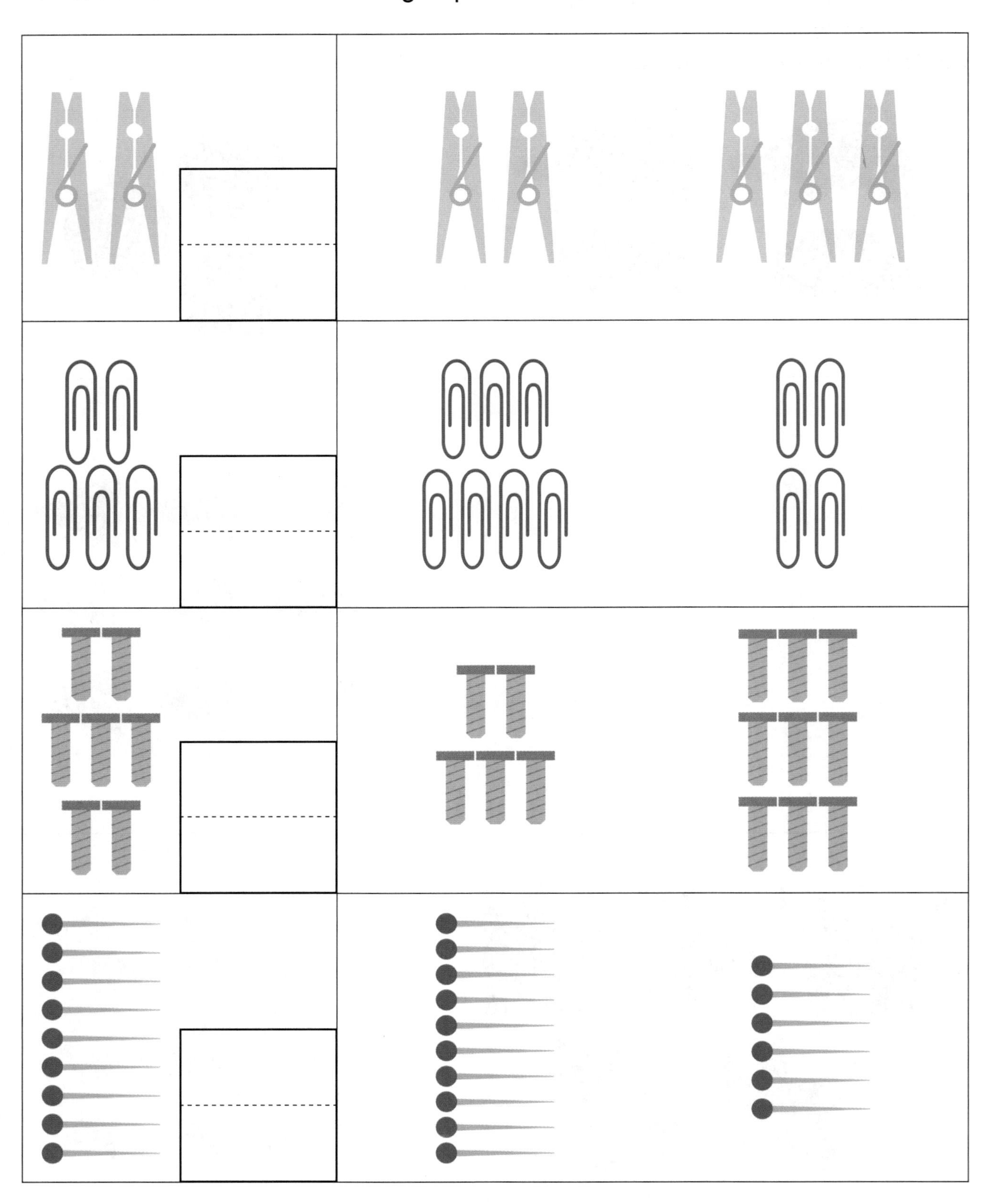

Using this page: Have students count the set on the left and write the number in the box, then count the other two sets and circle the set that has more.
Concept: Identifying the set that has more.

Circle the group that has fewer.

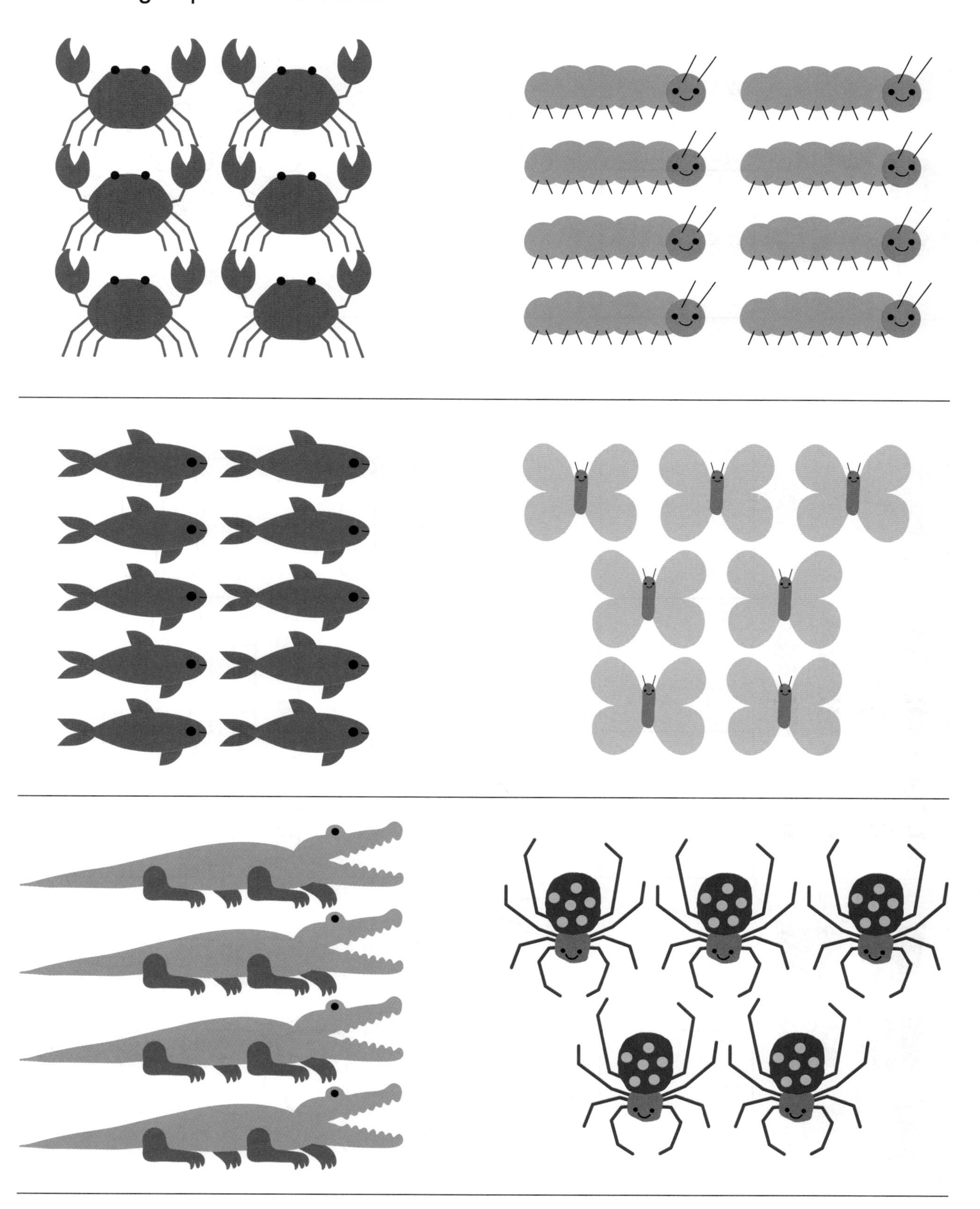

Using this page: Have students circle the set that has fewer in each row.
Concept: Identifying the set that has fewer.

Write the number and color the group that has fewer.

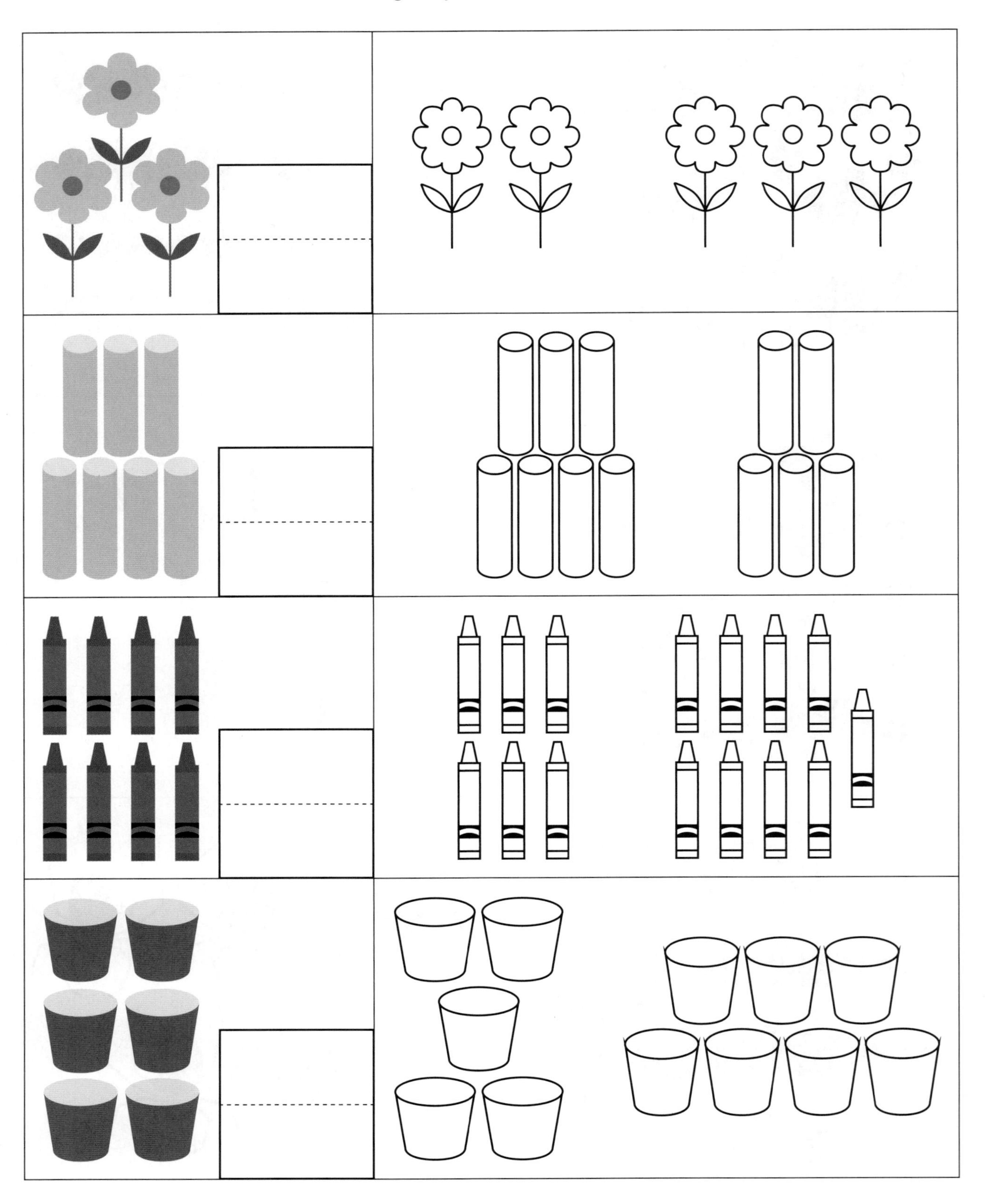

Using this page: Have students count the set on the left and write the numeral in the box, then count the other two sets and color the set that has fewer.
Concept: Identifying the set that has fewer.

Exercise 3

Compare and write the number.

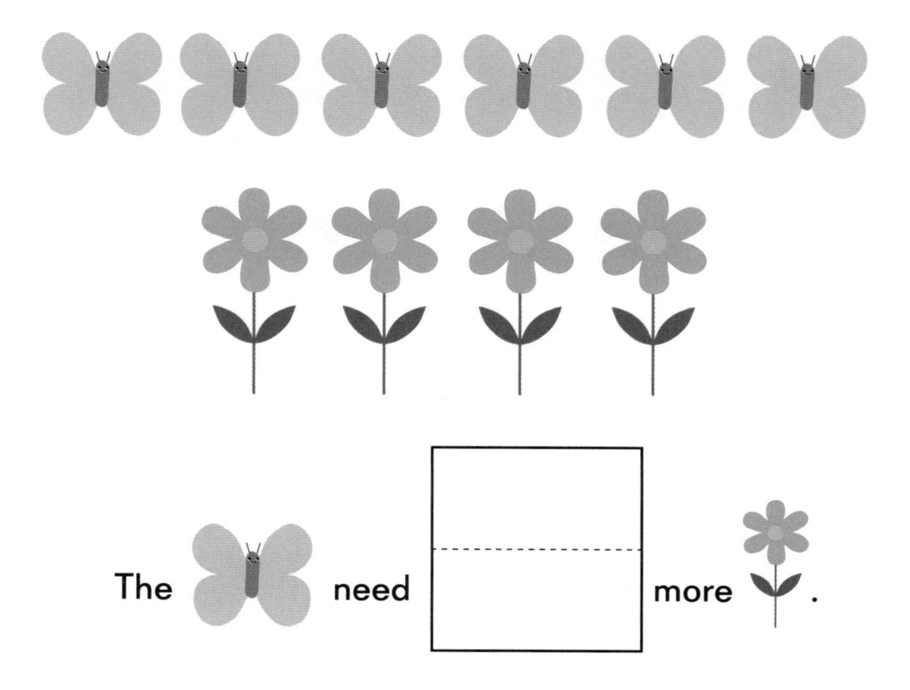

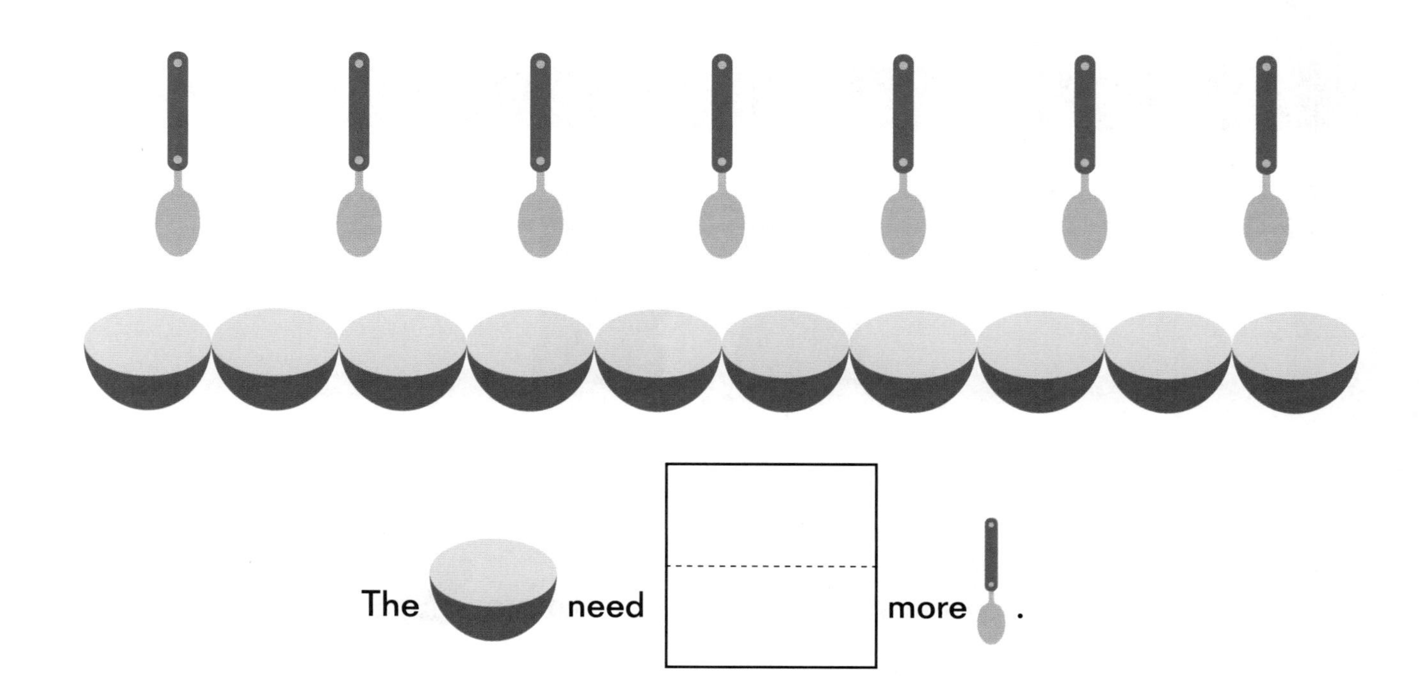

Using this page: Have students compare the two sets of objects, then write the numeral that shows how many more are needed to make both sets equal.
Concept: Comparing sets and finding the difference.

Compare and write the number.

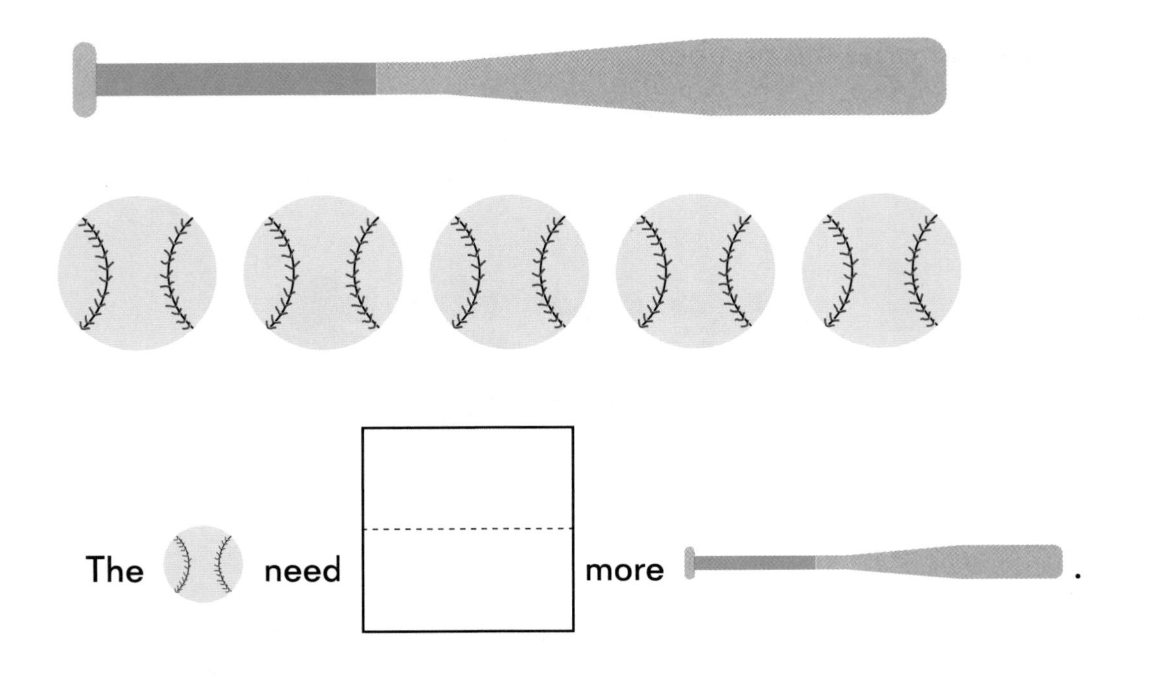

The ⚾ need ☐ more.

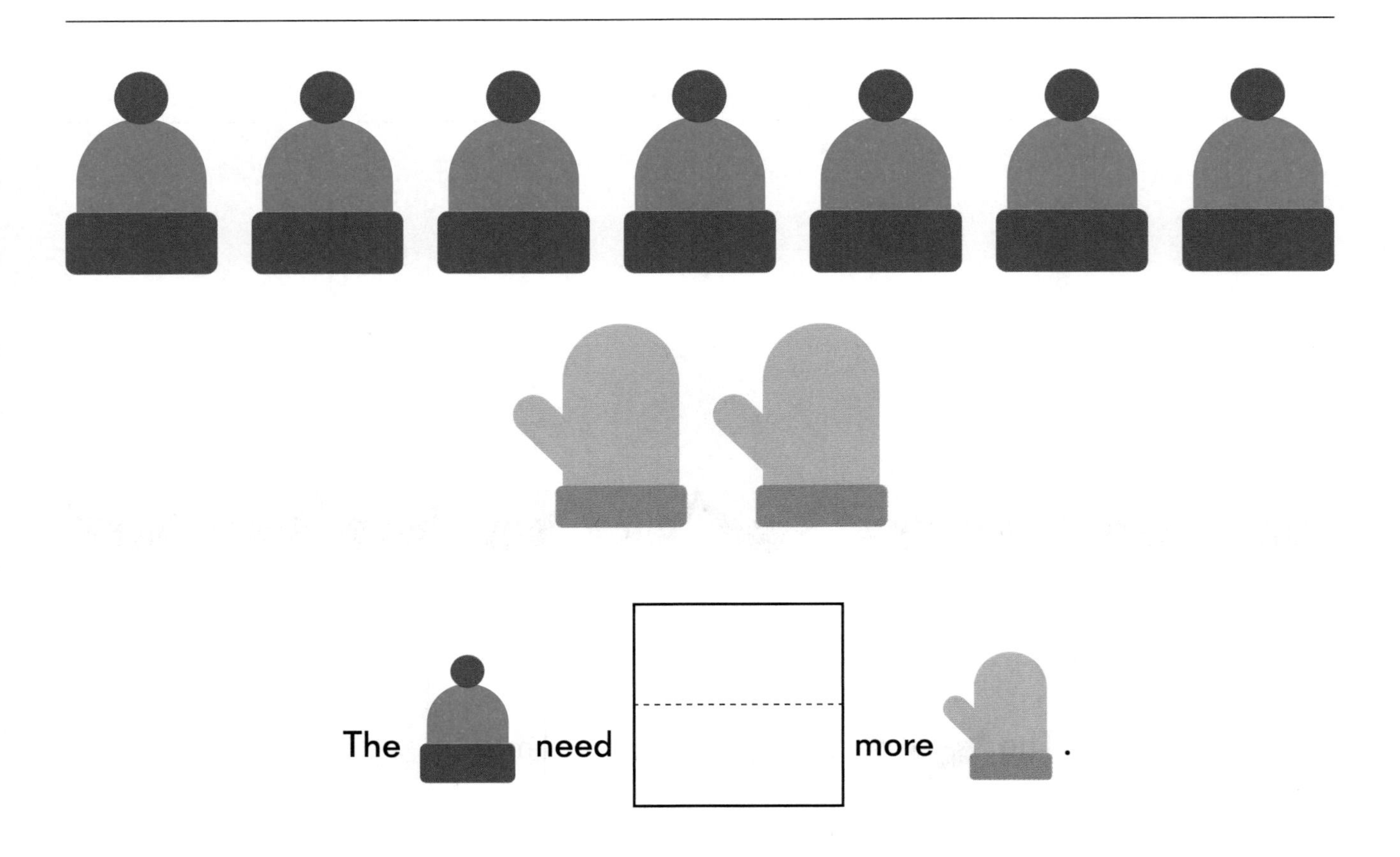

The need ☐ more.

Using this page: Have students compare the two sets of objects, then write the numeral that shows how many more are needed to make both sets equal.
Concept: Comparing sets and finding the difference.

Count and write the number.

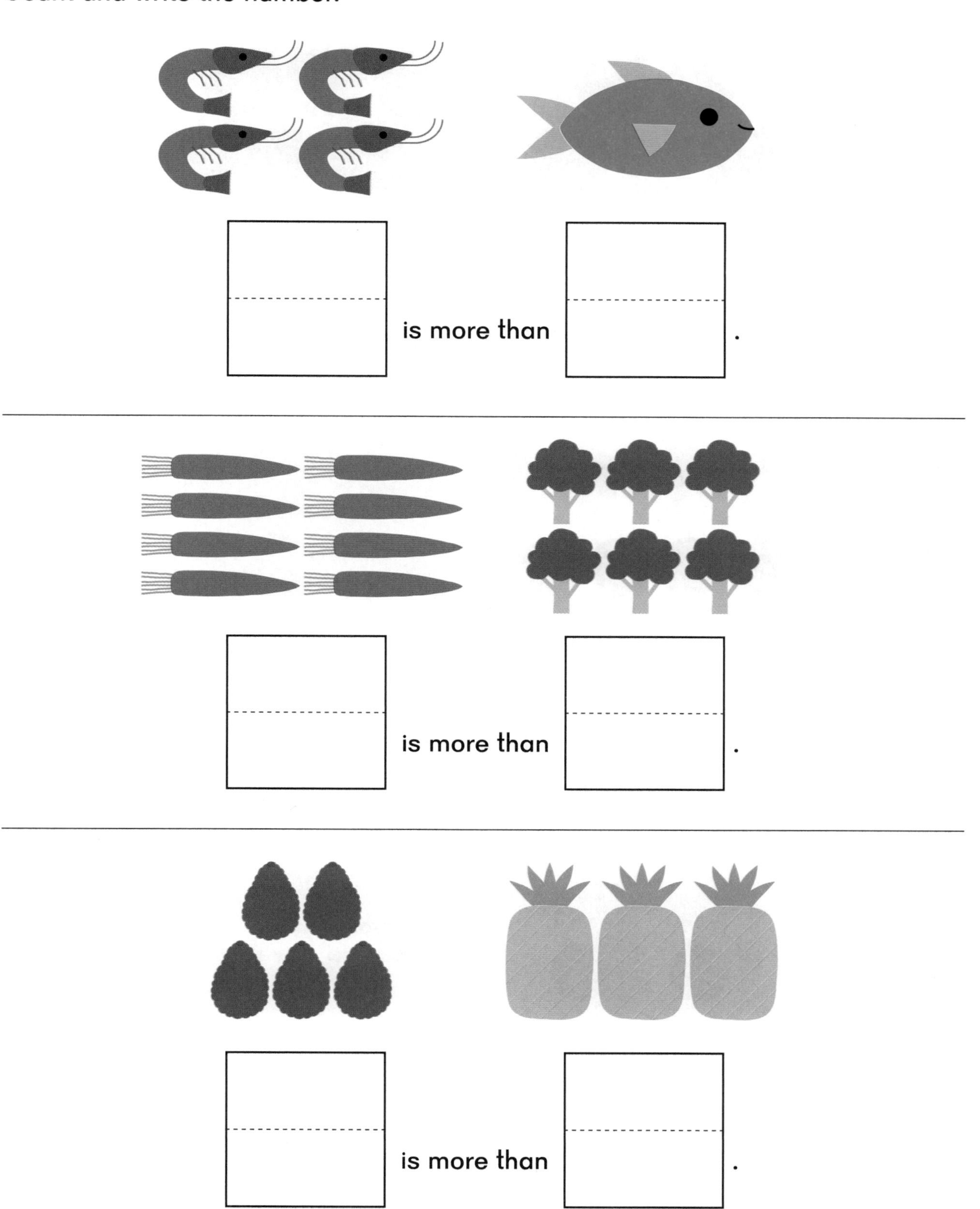

Using this page: Have students compare the sets of objects, then write the numerals in the boxes.
Concept: Comparing sets of objects using "more than."

Count and write the number.

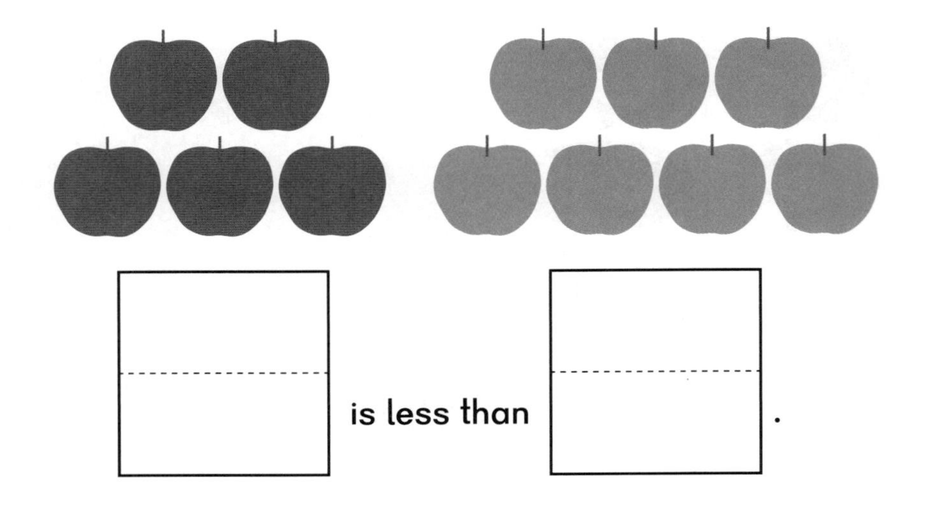

is less than .

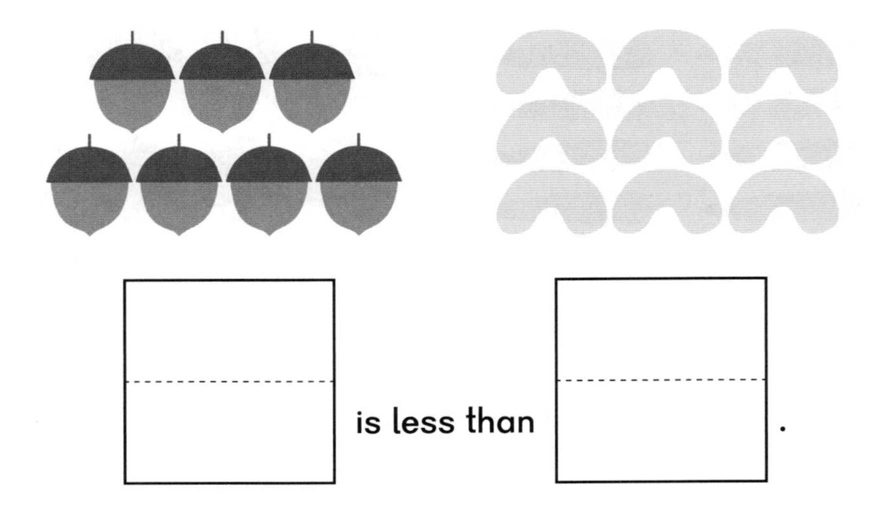

is less than .

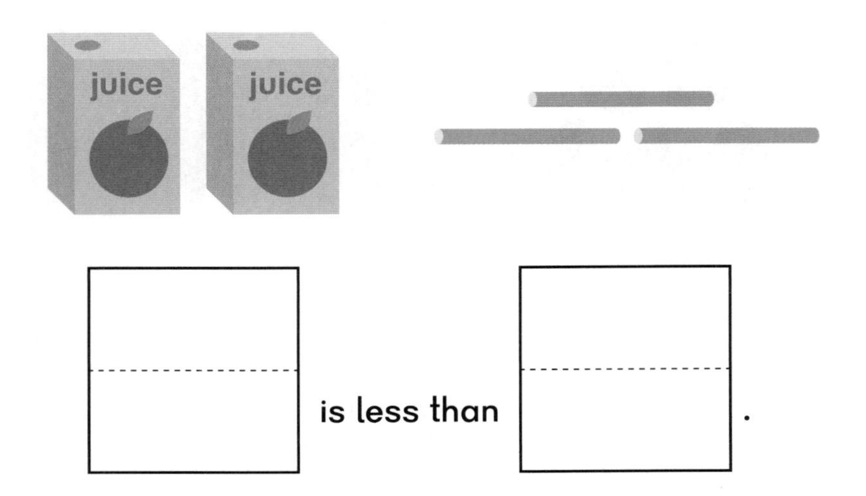

is less than .

Using this page: Have students compare the sets of objects, then write the numerals in the boxes.
Concept: Comparing sets of objects using "less than."

Exercise 4

Circle the sets that are equal.

Circle the set that has more.

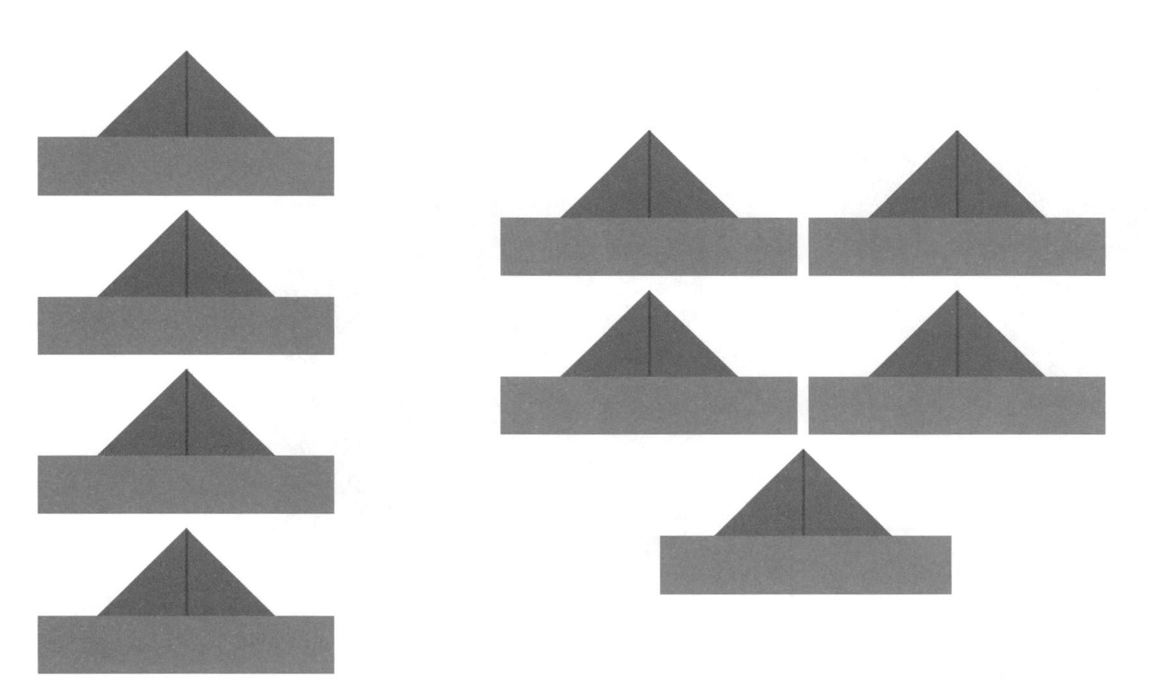

Using this page: Have students compare the sets of objects and circle the sets that are equal/the set that has more.

Write the number and circle the set that has fewer.

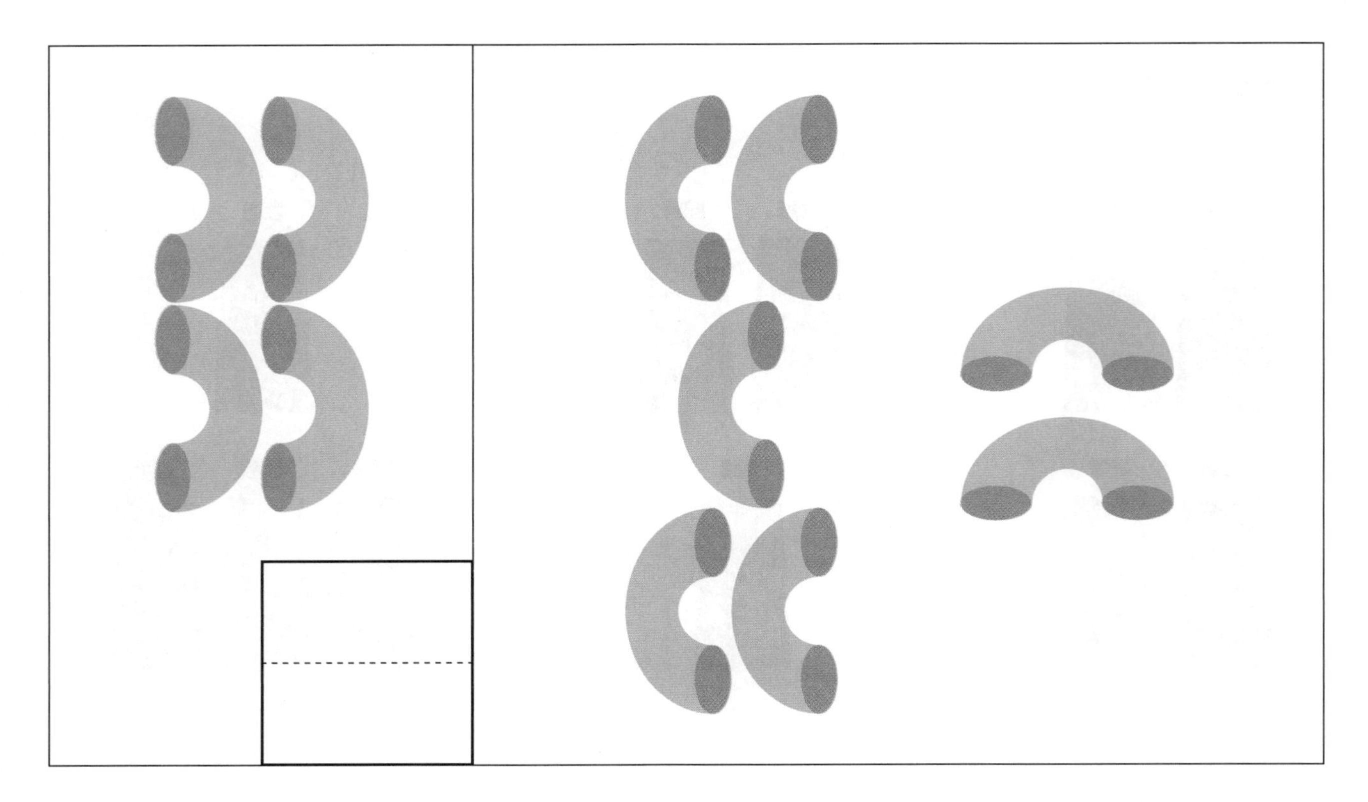

Write the number and circle the set that has more.

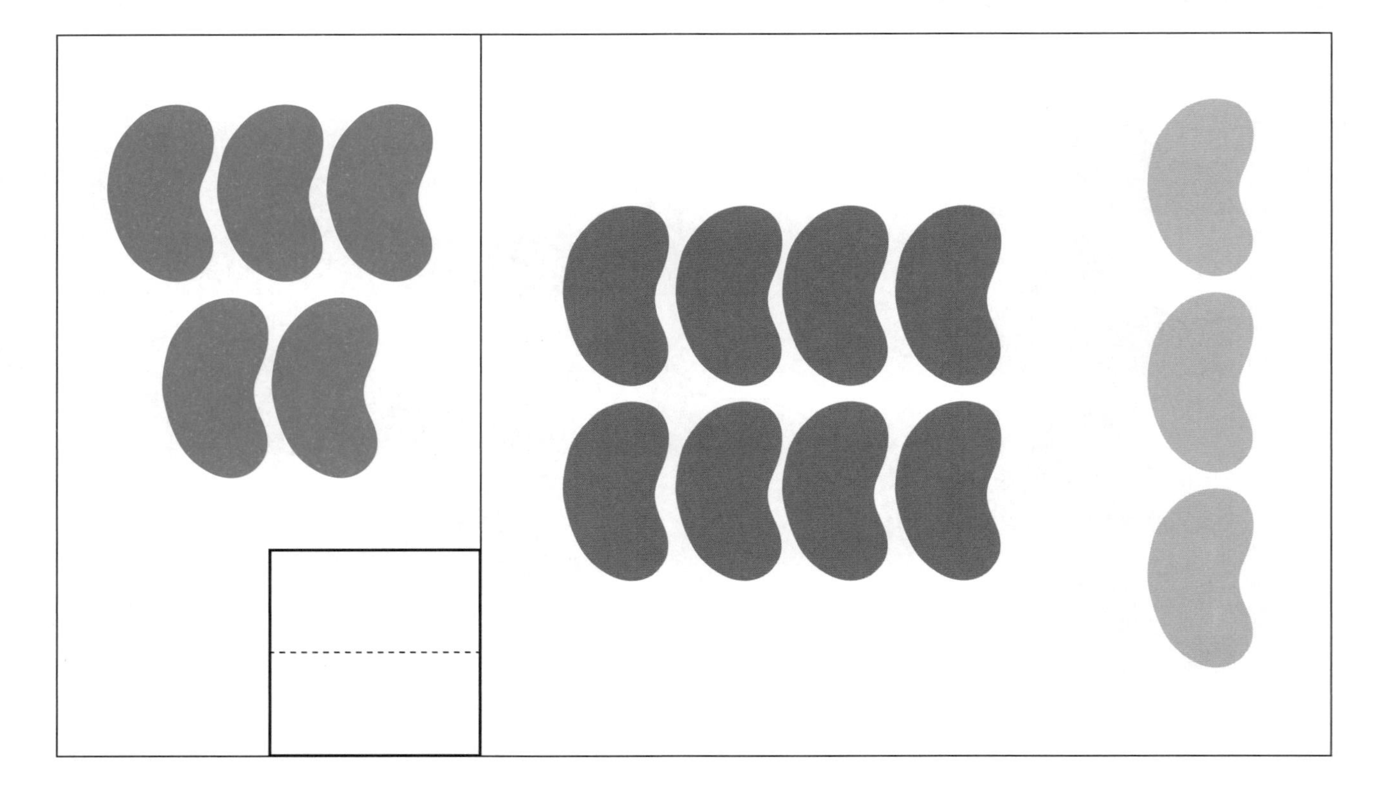

Using this page: Have students count the objects and write the numeral in the box, then have them circle the set that has fewer/more.

Exercise 5

Count and write the number.

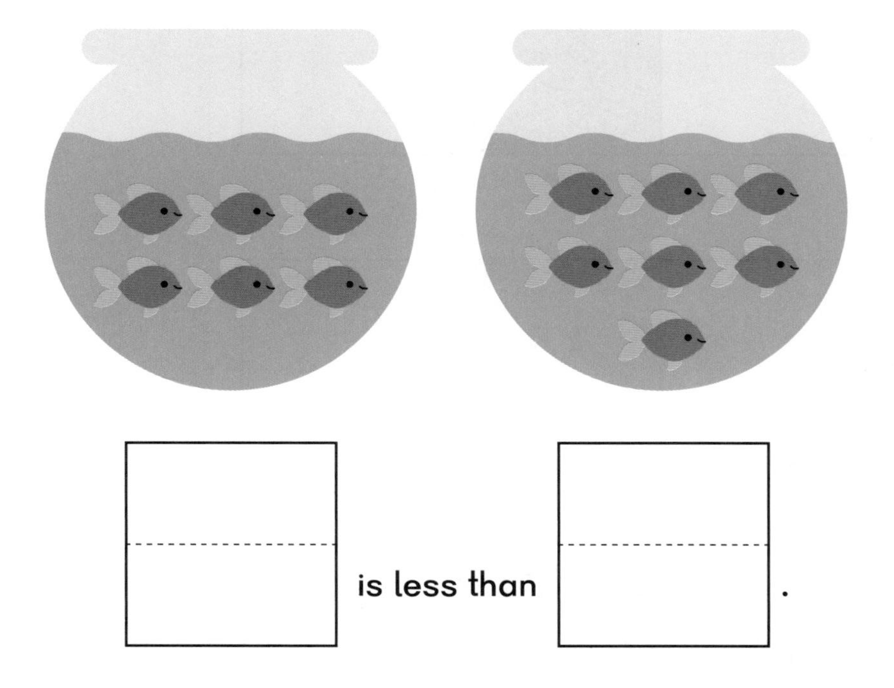

is less than .

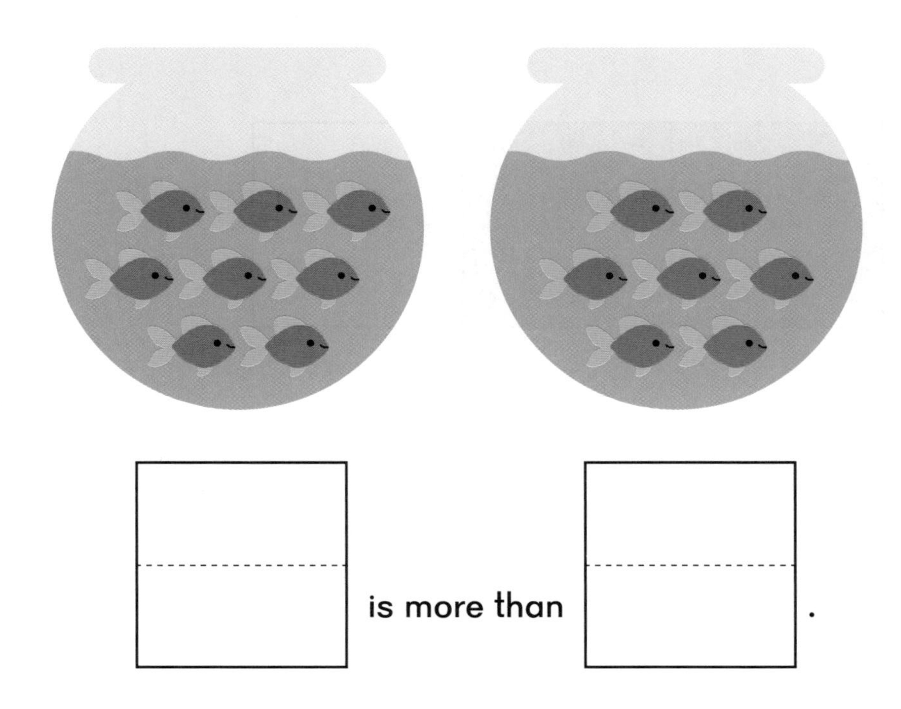

is more than .

Using this page: Have students count the fish in the bowls and write the numeral in the box. Then have them read the sentences.

Write the missing numbers.

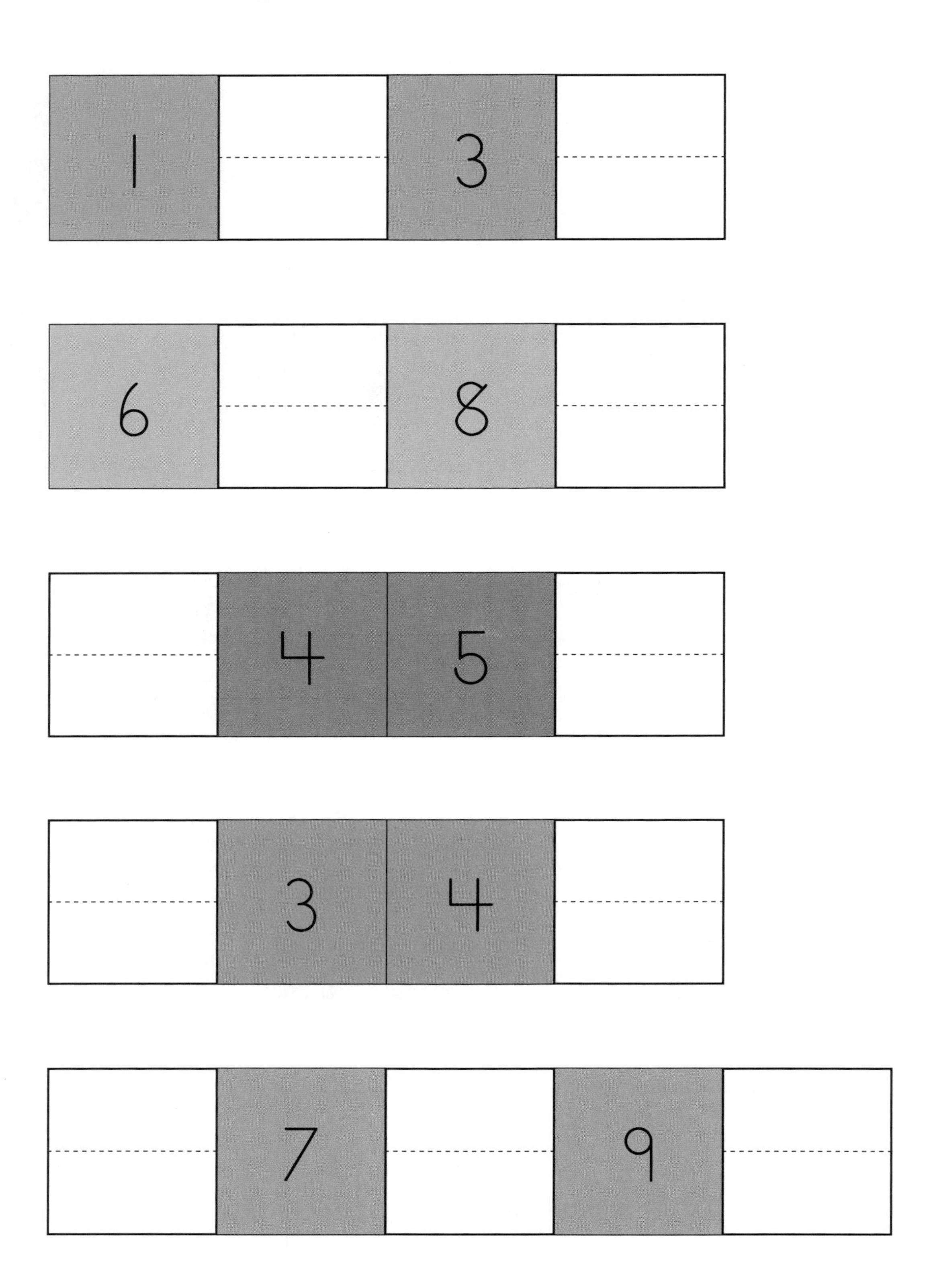

Using this page: Have students fill in each box with a numeral that is one more than the one before it and one less than the one after it. Then have them read the numerals in order.

Cut-outs for page 11

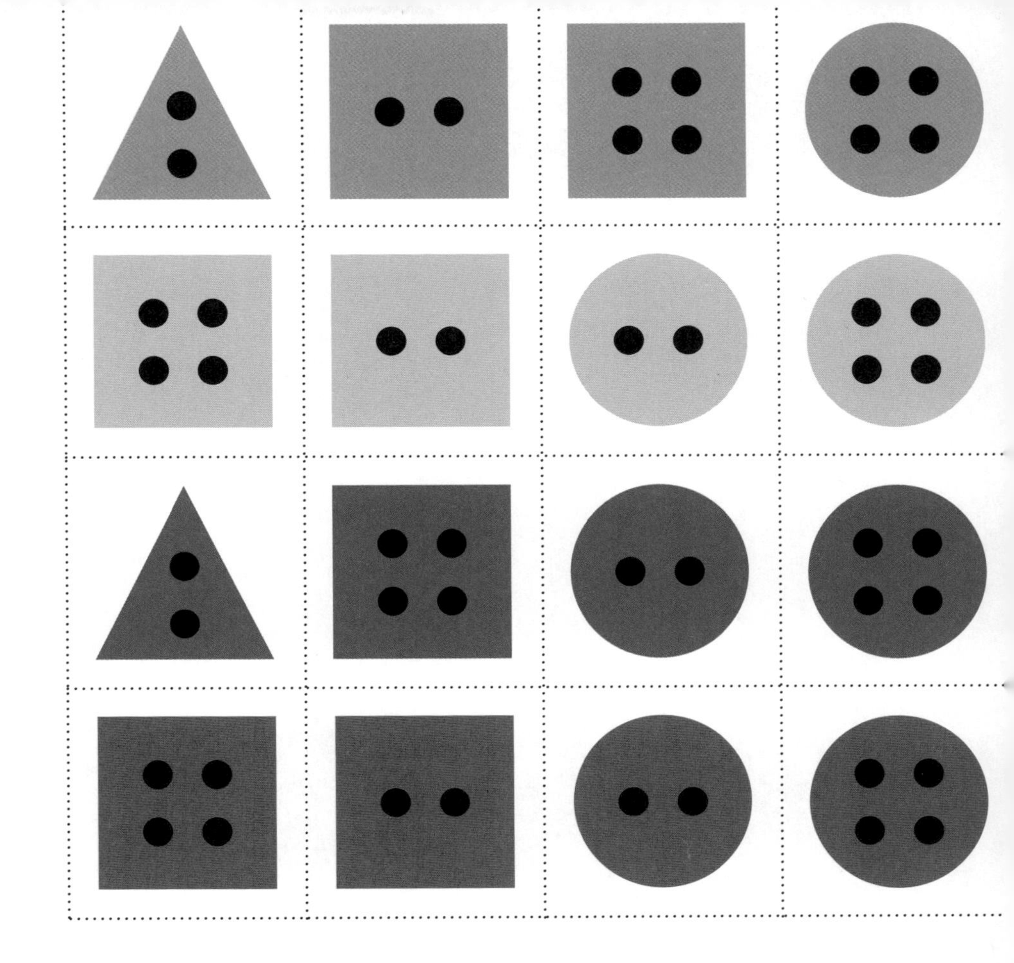

Cut-outs for page 12

Blank

Cut-outs for page 94

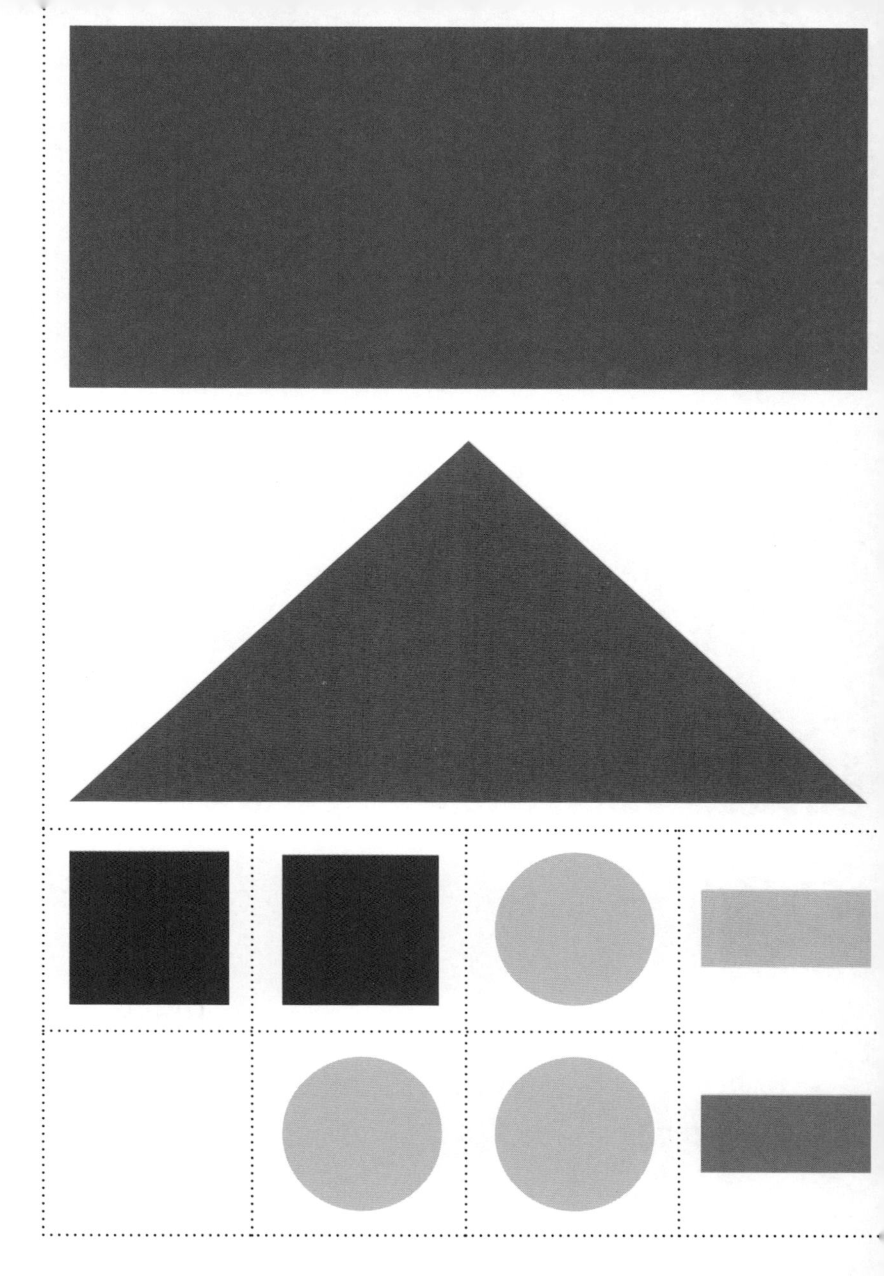

Blank

Cut-outs for page 100

Cut-outs for page 108

Cut-outs for page 110

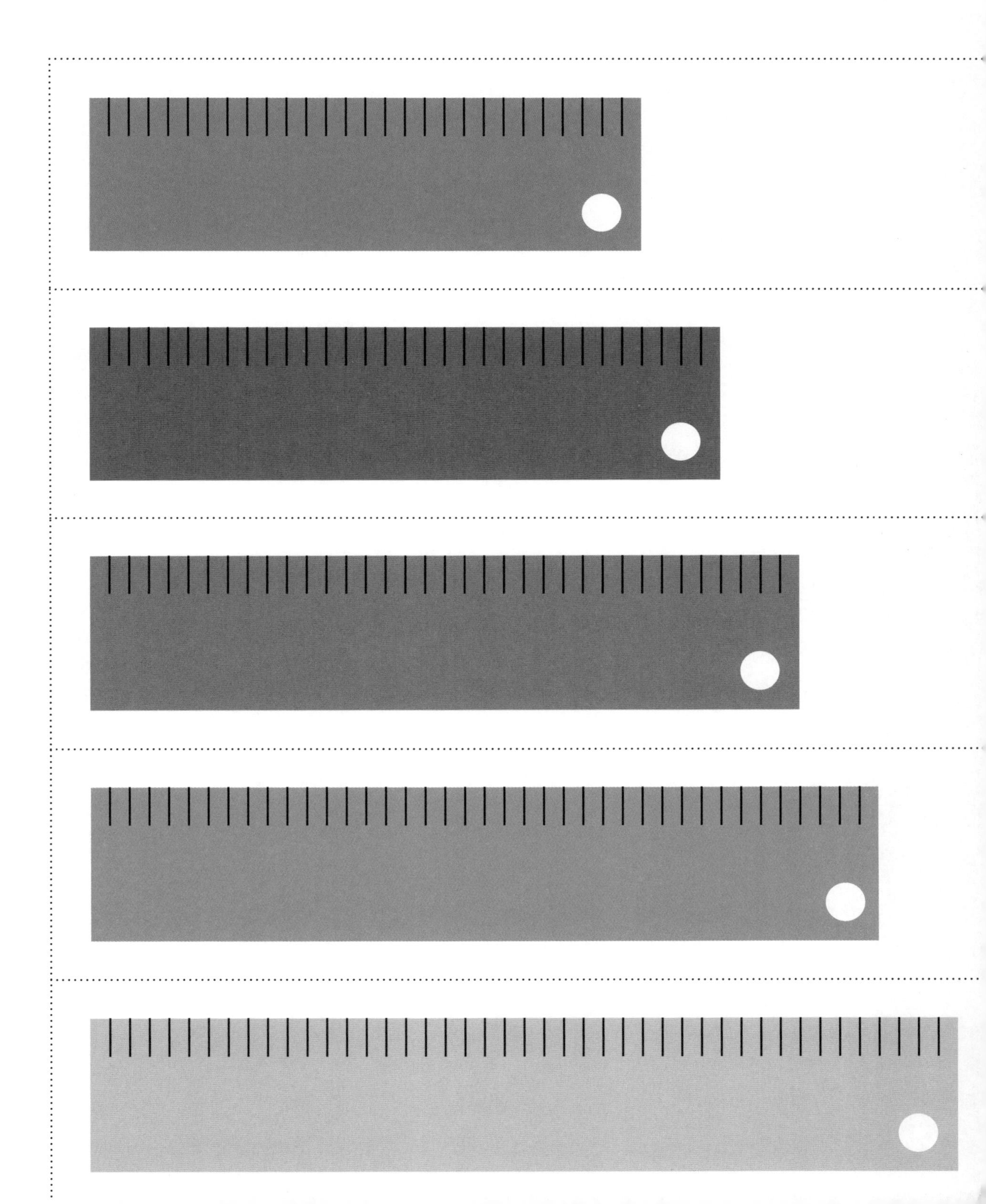